What It Means to Be 98% Chimpanzee

JONATHAN MARKS

What It Means to Be 98% Chimpanzee

Apes, People, and Their Genes

UNIVERSITY OF CALIFORNIA PRESS
BERKELEY   LOS ANGELES   LONDON

University of California Press
Berkeley and Los Angeles, California

University of California Press, Ltd.
London, England

First paperback printing 2003

Library of Congress Cataloging-in-Publication Data

Marks, Jonathan (Jonathan M.).
  What it means to be 98% chimpanzee : apes, people,
    and their genes / Jonathan Marks.
      p.   cm.
  Includes bibliographical references and index.
  ISBN 0-520-24064-2 (pbk : alk. paper)
  1. Human beings—Animal nature. 2. Human genetics.
  3. Human molecular genetics. 4. Human evolution.
  I. Title: What it means to be 98% chimpanzee.
  II. Title.

GN280.7 M37   2002
599.93′5—dc21   2001007085

Manufactured in the United States of America

11   10   09   08   07   06   05   04   03
10   9   8   7   6   5   4   3   2   1

For Peta and Abby

CONTENTS

List of Illustrations
xi

Acknowledgments
xiii

Preface to the Paperback Edition
xv

Introduction
I

ONE
MOLECULAR ANTHROPOLOGY
7

TWO
THE APE IN YOU
32

THREE
HOW PEOPLE DIFFER FROM ONE ANOTHER
51

FOUR

THE MEANING OF HUMAN VARIATION

72

FIVE

BEHAVIORAL GENETICS

100

SIX

FOLK HEREDITY

128

SEVEN

HUMAN NATURE

159

EIGHT

HUMAN RIGHTS...FOR APES?

180

NINE

A HUMAN GENE MUSEUM?

198

TEN

IDENTITY AND DESCENT

219

ELEVEN

IS BLOOD REALLY SO DAMN THICK?

242

TWELVE

SCIENCE, RELIGION, AND WORLDVIEW

266

Notes and Sources
289

Index
303

ILLUSTRATIONS

1. Tulp's "satyr," c. 1641.   16

2. Tyson's "orang-outang," or "pygmie," c. 1699.   18

3. A fusion of two ape chromosomes led to the formation
   of human chromosome 2.   38

4. Unlike humans, chimps and gorillas have darkly staining
   regions at the tips of their chromosomes.   39

5. Paraphyly: humans are phylogenetically apes only in the
   same sense that we are also phylogenetically fish.   46

6. Gorilla and man drawn by anatomist Adolph Schultz,
   1933.   71

7. The ambiguity of presumptive genetic relations among
   human "races."   201

ACKNOWLEDGMENTS

This book was begun in Berkeley and completed in Charlotte, and the research was supported in part by a grant from the National Science Foundation. It was a long time developing, and I'd like to thank the people who read and commented on it in various forms, in part or as a whole: Deborah Weiss, Ken Korey, Amos Deinard, Phyllis Dolhinow, Jonathan Beckwith, Alan Goodman, Jim Moore, Vincent Sarich, Phillip Johnson, Rika Kaestle, Debra Harry, Sarah Tishkoff, the participants in the 1999 Wenner-Gren symposium called "Anthropology in the Age of Genetics," and Regula Noetzli. Thanks as well to my editors at the University of California Press, Howard Boyer and Naomi Schneider.

PREFACE TO THE PAPERBACK EDITION

THE PAPERBACK EDITION OF *What It Means To Be 98% Chimpanzee* gives me a chance to rethink and update what I set out to do, which was to relativize the genetic place of humans and apes: not to deny it or challenge it, but simply to place that scientific work in an appropriate cultural and historical context, and to have explored its meaning. I still intend this book to be intensely scientific but also accessible, celebrating the best and abjuring the worst of science. Science is, at its best, self-correcting and relentlessly self-critical.

The issues I discuss remain as vital as ever. Among the new studies published since the book first appeared, some of the material exhibits the same old fallacies: confusing folk ideologies for natural patterns, humanizing apes and ape-ifying humans, and under-theorizing genetic data in an anthropological vacuum. But much of this material, I'm happy to say, is good, normal science. Here are some summaries:

• Svante Pääbo and his colleagues at the Max Planck Institute for Evolutionary Anthropology have found that patterns of gene expression in humans and chimps are quite a bit different in the brain, but less different than in other organs—as we might well expect.

• The biochemist Roy Britten argues that the genetic similarity between humans and chimpanzees is not over 98%, but closer to

95%, when you count insertions and deletions of DNA (chapter two). Unfortunately, the problem is not the actual number, but rather the fact that the diversity of genomic mutational processes precludes the calculation of a single value for the genetic difference between human and chimpanzee with any degree of rigor.

• The search for genetic changes that translate directly into the physical distinctions between human and ape is likely futile, but undertaking such a search is tempting nevertheless (chapter eleven). In the quest for what makes us human, differences have been now found between the human and chimpanzee versions of a "brain gene" (CMAH) and a "language gene" (FOXP2). Given that small differences between human and chimpanzee genes will be found everywhere, and that the most basic fact of modern genetics is that phenotypes do not map readily onto genotypes, this kind of theorizing seems conceptually naive, even if high-tech.

• The molecular evidence for the domestication of the dog (chapter four) has now been revised to bring it more in line with the archaeological evidence, and the sequences of Neandertal DNA, while still a technical triumph, show less about the relationships of humans and Neandertals than was initially thought.

• Orangutans now are recognized to possess shared, learned behaviors—as chimpanzees do (chapter eight). How could animals for whom maturity is so delayed and learning is so important to their survival be any different? But it is still not clear how relevant that is to what anthropologists call "culture" in humans. Labeling ape behavior as "culture" simply means you have to find a different word for what humans do.

• The Human Genome Diversity Project (chapter nine) now uses a European collection of cells from indigenous peoples. Examination of these data still shows polymorphism and clinal variation to be the major patterns of human genetic diversity and suggest that geographic separation is the major determinant of the genetic differences among human groups. When researchers asked a new computer program to divide the world's genetic variation into two groups, the result broadly separated Europe, Africa, and Western Asia from the Far East, Oceania, and the New World. When asked to divide the world into five, the computer gave results corresponding to the peoples of Africa, Eurasia, East Asia, Oceania, and the Americas; and when asked to divide the world into six, it separated out the Kalash people of Pakistan from those five. On the face of it, this finding would seem to lend no support to popular ideas about race: the Kalash are hardly equivalent to the Africans; and there is certainly nothing racially commonsensical about juxtaposing Eurasia and Africa against East Asia, Oceania, and the Americas. How the *New York Times* came to represent these newest findings under the headline "Gene Study Identifies 5 Main Human Populations" in their issue of December 20, 2002—along with the ridiculous assertion that they "essentially confirmed the popular conception of race"—is anybody's guess.

• Other population geneticists have recently argued that human races really do exist as natural units, in spite of much of the last half century of research in anthropology and genetics (chapter four). They do so partly by perversely using "population," "race," and "ethnic group" synonymously—although the first term is supposed to be locally, the second globally, and the third culturally designated. If you fail to acknowledge that human groups are fluid, hierarchically organized, and symbolically bounded, it's hard to imagine that the science you do will be of much value. A brief paper by the bioethicists Pamela Sankar and Mildred Cho has effectively responded to and explored the ambiguities inherent in scientific research that uses race as if it were a natural category. The epidemiologists Jay Kaufman and Susan Hall brilliantly dispose of the scientific canard that hypertension in blacks is a genetic disease. And the sociologist Troy Duster has brought to light the financial interests of racialized pharmacogenetics in some of the newest contentious claims.

• A company in Florida will sell you a genetic test to tell you what race you are (as if you don't already know). And then they'll tell you what percentage of each racial ancestry you have. When you're done, be sure to visit the pet psychic. *Caveat emptor!*

• Judge John Jelderks has been consistently ruling against the Native American tribes in the case of Kennewick Man (chapter ten), but science still hasn't found anything about him that it didn't already know, and the whole affair has left a bad taste in many people's mouths. One of the major scientific plaintiffs in the case has even published a paper attacking, of all people, the anthropologist Franz Boas—claiming that Boas's original immigrant study from 1910, which spawned the field of human adaptability studies, did not really show any change in cranial form, which is fixed genetically and racially. That assertion has the ring of desperation to it (if not obfuscation), and has been quickly rebutted by more mainstream biological anthropology.

• We have heard about the skull of "Peñon Woman"—13,000 years old and looking like the skull of Japanese Ainus, leading a researcher in England to crow, "We're going to say to Native Americans, 'Maybe there were some people in the Americas before you, who are not related to you.'" Silly me, I thought we were *all* related.

• Peñon Woman is also claimed by the scientists who push a pre-Indian European presence in the New World. But there are still voices of reason amid the chaos: "I personally haven't found it very convincing," said the anthropologist Chris Stringer to a London newspaper. "For a start, there are lots of examples in archaeology where various artifacts from different parts of the world can end up looking similar even though they have different origins . . . Most humans in the world at that time were long headed and it doesn't surprise me that Peñon woman at 13,000 years old is also long headed."

• And with a rhetorical flourish, a new study from Morris Goodman's group proclaims (again) that "DNA evidence provides an objective non-anthropocentric view of the place of humans in evolution. We humans appear as only slightly remodeled chimpanzee-like apes." Here "objective" seems to denote merely a blindness to your own intellectual prejudices; for the "slightly remodeled" DNA still produces two species which are adaptively and ecologically rather distinct from each other—incorporating into one of them such trivialities as walking, talking, sweating, sensuality, and civilization (chapter two). That the molecular geneticist can't tell the human from the ape doesn't mean that nobody can; and the sooner they confront that fact, the better.

• All discussions of apes end with conservation. A recent survey finds that there are fewer than half as many chimpanzees and gorillas in Gabon today as there were twenty years ago. The principal causes of this decline in numbers are deforestation, hunting, and disease. While I maintain that overstating the humanness of the apes is not a valid justification for our attitudes and actions, it is nevertheless also the case that our natural and cognitive worlds would be immeasurably impoverished by the loss of these species. Broad international action to conserve the apes is desperately needed, in conjunction with efforts to provide for the needs of the local people.

It now seems to me that there are seven well-established patterns of human genetic variation, understanding the meaning of which will be the challenge of the new molecular anthropology:

1. Humans and apes are very similar genetically.

2. They are so similar that different genetic data sets often tell conflicting phylogenetic stories.

3. Compared to apes, humans are depauperate in genetic variation.

4. Human genetic variation is principally polymorphic.

5. Human genetic variation between groups is clinal.

6. All people are genetic subsets of Africans.

7. Genetic variation and behavioral variation do not map easily on to one another in the human species.

I thank Alan Vanneman for correcting my memory about Sherlock Holmes.

Jonathan Marks
Charlotte, North Carolina
April 2003

REFERENCES

Britten, R. J. "Divergence between Samples of Chimpanzee and Human DNA Sequences Is 5%, Counting Indels." *Proceedings of the National Academy of Sciences, USA* 99 (2002): 13633–35.

Connor, S. "Does Skull Prove That the First Americans Came from Europe?" *Independent* (London), December 4, 2002.

Cooper, R. S., J. S. Kaufman, and R. Ward. "Race and Genomics." *New England Journal of Medicine* 348 (2003): 1166–70.

Duster, T. "Medicine and People of Color." *San Francisco Chronicle*, March 17, 2003.

Enard, W., P. Khaitovich, J. Klose, S. Zollner, F. Hessig, P Giavalisco, K. Nieselt-Streuwe, E. Muchmore, A. Varki, R. Ravid, and S. Pääbo. "Intra- and Interspecific Variation in Primate Gene Expression Patterns. *Science* 296 (2003): 340–43.

Gravlee, C., H. Berhard, and W. Leonard. "Heredity, Environment, and Cranial Form: A Reanalysis of Boas's Immigrant Data." *American Anthropologist* 105 (2003): 125–38.

Hagelberg, E. "Recombination or Mutation Rate Heterogeneity? Implications for Mitochondrial Eve." *Trends in Genetics* 19 (2003): 84–90.

Kaufman, J., and S. Hall, S. "The Slavery Hypertension Hypothesis: Dissemination and Appeal of a Modern Race Theory." *Epidemiology and Society* 14 (2003): 111–26.

Legon, Jeordan 2002. "Scientist: Oldest American Skull Found." www.cnn.com/2002/TECH/science/12/03/oldest.skull/index.html. Retrieved May 7, 2003.

Leonard, J. A., R. K. Wayne, J. Wheeler, R. Valadedez, S. Guillen, and C. Vila. "Ancient DNA Evidence for Old World Origin of New World Dogs." *Science* 298 (2002): 1613–16.

Pääbo, S. "The Mosaic That Is Our Genome." *Nature* 421(2003): 409–11.

Relethford, J. "Absence of Regional Affinities of Neandertal DNA with Living Humans Does Not Reject Multiregional Evolution." *American Journal of Physical Anthropology* 115 (2001): 95–98.

Risch, N., E. Burchard, E. Ziv, and H. Tang. "Categorization of Humans in Biomedical Research: Genes, Race, and Disease." *Genome Biology* 3 (2002): 2007.1–12.

Rosenberg, N. A., J. K. Pritchard, J. L. Weber, H. M. Cann, K. K. Kidd, L. A. Zhivotovsky, and M. W. Feldman. "Genetic Structure of Human Populations." *Science,* 298 (2002): 2181–85.

Sankar, P., and M. K. Cho. "Toward a New Vocabulary of Human Genetic Diversity." *Science* 298 (2002): 1337–38.

Sparks, C. S., and R. L. Jantz. "A Reassessment of Human Cranial Plasticity: Boas Revisited." *Proceedings of the National Academy of Sciences, USA* 99 (2002): 14636–39.

Wade, N. "For Sale: A DNA Test to Measure Racial Mix." *The New York Times,* October 1, 2002.

Walsh, P. D., et al. "Catastrophic Ape Decline in Western Equatorial Africa." *Nature* 422 (2003): 611–14.

Wildman, D. E., M. Uddin, Guozhen Liu, L. Grossman, and M. Goodman. "Implications of natural selection in shaping 99.4% nonsynonymous DNA identity between humans and chimpanzees: Enlarging genus *Homo*." *Proceedings of the National Academy of Sciences, USA* 100 (2003): 7181–7188.

INTRODUCTION

C. P. SNOW, WHO WAS BOTH a scientist and a novelist, observed in a classic essay from the 1950s that the sciences and the humanities were coming apart at their academic seams and forming "two cultures." He meant this in a specifically anthropological sense—two communities that speak different languages, see the world in different ways, don't understand each other, and regard each other with suspicion. Each thinks itself superior to the other.

This rift is probably irreparable. As the frontiers of knowledge have expanded, it has become hard enough to keep up with the work in one's own ridiculously narrow field of expertise, never mind to read novels, philosophy, or particle physics besides. Scientists lament the lack of science education on the part of the public, but they have ceded to science journalists the responsibility of educating the public. Humanists lecture about the construction of knowledge, but scientists lecture that they are simply recording what is "out there."

This book is about a hybrid field that we can call "molecular anthropology." To a large extent, it epitomizes the insecurities of modern science. On the one hand, technology permits us to study aspects of the human condition in far greater detail than was previously thought possible—that's the meaning here of "molecular." On the other hand, the scientists themselves have often employed that information to prop up dubious political assertions; or else they have interpreted the information through cultural lenses of various tints, and often with striking naïveté—that's the "anthropology."

Technical sophistication and intellectual naïveté have been the twin hallmarks of human genetics since its origins as a science in the early part of the twentieth century. The way genetics was practiced and preached in the 1920s exploited the cachet of modern science to justify blatant racism and xenophobia.

Times have changed, and technologies have certainly changed. But many of our cultural ideas have remained strikingly unaltered across the generations. We have a strong faith in the power of heredity to shape destiny, in the ability of modern science to arrive at truths about nature, in our identity as a deeply inscribed property, in the constitution of scientific facts to be neither good nor bad (but just authoritative), and in the ability of those scientific facts to speak for themselves.

Each of those propositions is true only to a very limited extent. What is needed in human genetics is a mediation of its fundamentally scientific and humanistic elements.

Anthropology has always been a field of mediation. Classically (in the 1920s), it involved juxtaposing the exotic and the mundane—showing that your way of seeing and interpreting the world is only one of many possible and valid ways, but at the same time showing that what New Guinea tribesmen do is only superficially different from what you do.

In more recent decades, anthropology has assumed the political role of mediator for aboriginal populations (usually the objects of anthropological study, of course) and colonial powers (usually the ones sending the anthropologist out). On the biological end, anthropology emphasizes, on the one hand, the continuity of humans with other primates, but, on the other, the uniqueness of humans among the primates. And in a more general sense, anthropology mediates between professional scholarly knowledge about the world ("science") and popular or cultural wisdom about it ("folk knowledge").

Molecular anthropology necessarily adopts the crucial role of mediator as well. Genetics advertises a classically modern scientific analysis of the human condition, and thus molecular anthropology examines both human populations with respect to one another and our species with respect to other species. At the same time, however, we are forced to ask what meaning to attach to such studies and what value they have. Where human lives, welfare, and rights are concerned, genetics has historically provided excuses for those who wish to make other people's lives miserable, to justify their subjugation, or to curry favor with the wealthy and powerful by scapegoating the poor and voiceless. It is therefore now obliged to endure considerably higher levels of scrutiny than other, more benign and less corruptible, kinds of scientific pronouncements might.

Rather than simply avowing to study our hereditary constitution objectively, dispassionately, and benignly—and being proved wrong time and again—this book is about the way a genetic science of humanity can confront issues. Some of these issues are political, such as animal rights and colonialism; others lie in the domain of folk wisdom, such as ethnocentrism and racism; and still others lie in simply the way science represents itself to the public.

"Molecular anthropology" is a term paradoxically coined by a biochemist in 1962 to designate the study of human evolution by recourse to the differences in the structure of biomolecules. The paradox is that although it sounds like a kind of anthropology, a *molecular* kind of anthropology, it was really the technology of biochemistry merely being applied to classically anthropological questions. And since technology drove this new field, anybody could do "molecular anthropology," regardless of how much anthropology they really knew.

While that may sound harmless enough, consider the opposite case. What would constitute an "anthropological biochemistry" if you didn't need to know any biochemistry to do it?

What I will show in this book is that when the cutting-edge technology of molecular genetics has been wed to a "folk knowledge" of anthropology, the results have invariably been of exceedingly limited value. This was true in the 1920s, when geneticists sought to rewrite our understanding of social issues by blaming poverty on the genes of the poor. The stock market crash and Depression had a sobering effect on the geneticists.

It was also true in the 1960s, when genetics became molecular and its practitioners began to make observations of seeming profundity, such as "from the standpoint of hemoglobin, man is just an abnormal gorilla." It seems not to have occurred to the sanguine speaker that the standpoint of hemoglobin might just be a poor one for the problem at hand: from the top of the Empire State Building, Chicago and Los Angeles appear to be in the same place over the horizon. But not from the Golden Gate Bridge.

That's a classic anthropological question—whose standpoint is superior? An anthropological approach would be to inquire what it is that each standpoint allows you to see that the others conceal.

The standpoint of science is widely held to be superior to all rivals. Especially by scientists. But once again, it is useful to acknowledge that there may be more than one scientific standpoint, and that the meaning of any particular scientific pronouncement may not be self-evident. And thus in the 1990s, we routinely heard that we are just 1 or 2% different from chimpanzees genetically, and therefore . . . what?

Should we accord the chimpanzees human rights, as some activists have suggested?

Should we acknowledge and accept as natural the promiscuity and genocidal violence that lurks just underneath the veneer of humanity and occasionally surfaces, as some biologists have implied?

Or should we perhaps all simply go naked and sleep in trees as the chimpanzees do?

None of these suggestions, of course, necessarily follows from the genetic similarity of humans to apes, although the first two have been proposed within the academic community and promoted in the popular media over the past few years. (Mercifully, the third has not.) But all of them *sound* as though they might well proceed from that genetic similarity.

An anthropological or cultural perspective allows us to examine critically some of the assumptions that we often take for granted about genetics itself.

The first topic this book addresses is: What does the genetic similarity of humans to apes mean? What is it based on? Does it have profound implications for understanding our nature?

Here we will see that the universe of genetic similarities is quite different from our preconceptions of what similarities mean. For example, the very structure of DNA compels it to be no more than 75% different, no matter how diverse the species being compared are. Yet the fact that our DNA is more than 25% similar to a dandelion's does not imply that we are over one-quarter dandelion—even if the latter were a sensible statement. This will be a primary illustration of the confrontation between scientific data and folk knowledge, and of the exploitation of the latter by the former. The extent to which our DNA resembles an ape's predicts nothing about our general similarity to apes, much less about any moral or political consequences arising from it.

From there, I go on to examine the genetic differences within the human species and how they have intersected with our attempts to classify people into races. Geneticists have attempted to track the evolutionary history of our species with varying degrees of success, often finding what they expect—identifying races in one generation, denying their existence in another. The perspective of molecular anthropology—a social science of heredity—will shed light on both the science itself and the uses of the science.

Perhaps the most contentious issue in modern biology, rekindled by the furor over Richard Herrnstein and Charles Murray's 1994 book *The Bell Curve,* is behavioral genetics. Here the pattern of human behavioral diversity can be compared to the known patterns of genetic variation, enabling us to look critically at the political claims ostensibly derived from the science.

Two modern social projects hoping to justify their existence by recourse to genetics are the Great Ape Project, which argues for human rights for apes on the grounds of our genetic kinship with them, and the Human Genome Diversity Project, which has advocated the establishment of a genetic museum of the isolated and endangered peoples of the world. Both of these proposals can be illuminated by bringing together the scientific and humanistic elements that bear upon them.

Finally, I explore more generally the ways in which technical and cultural knowledge intersect in the classic conflict between science and religion. This broadens our scope from a humanistic study of heredity to a culturally informed and socially relevant study of the role of science.

Ultimately, that is what molecular anthropology is all about: the intersection of chemical bodies, human bodies, and bodies of knowledge; and their mutual illumination. Molecular anthropology acts as mediator between reductive genetics and holistic anthropology; between formal knowledge and ideology; between facts of nature and facts produced by authorities; between what science can do and what scientists ought to do; and most fundamentally, between human and animal. All of these terms are, of course, laden with meanings, and none of them can be taken at face value.

That's the fun of it.

One

You know them. You've seen them, perhaps in the zoo, perhaps in the movies or on television. You've looked deep into their dark, soulful eyes, pondered their hairy faces, and recognized a mind behind those eyes. A mind like a child's perhaps, but a mind akin to your own. You've seen their sinewy arms, their long fingers clasping something—a twig to use as a tool? Or perhaps a Raggedy-Ann doll?

And you know about those resemblances. That chimpanzee, says the narrator, is more than 98% genetically identical to us. That similarity, he says, blurs the line between us. We are chimpanzees, and they are us.

But take another look. Those deep eyes—without whites, which those chimpanzee eyes really do lack—they look more like a dog's dark eyes than like your own, don't they? And those sinewy arms are rather hairy, aren't they? They aren't human arms at all—they are *like* human arms. The chimpanzees themselves could not really be confused with people. Their heads are smaller. Their eyeteeth are bigger. They have neither noses nor foreheads to speak of. Their legs are short, their ears are huge, and they use their hands in walking. They have thumbs on their feet. They have knees pointing ridiculously far outward.

Listen to them. You hear no speech; rather, a series of grunts and hoots. Sounds, meaningful sounds, but hardly human sounds.

They don't gossip, pray, sing, praise, or insult. They don't decorate their bodies or cover them up. Their sexuality is stimulated by colorful swelling of the female genitalia. In fact, the chimpanzees do hardly anything you might recognize as being human. When they do, after all, it's newsworthy.

And not only that, but sometimes the adults kill and eat the chimpanzee babies.

All of which is certainly not to deny that chimpanzees are closely related to us. In the panoply of nature, chimpanzees are very similar to us. Our species gives birth to live young and nurses them, like chimpanzees and other mammals and unlike salamanders and pigeons. Among the mammals, our species has grasping hands, toenails, and only one pair of nipples—a combination again like chimpanzees, but unlike cows or dogs or dolphins. And among the primates, our species has a very mobile shoulder and no tail—once again like chimpanzees, but unlike the awfully cute lorises and vervet monkeys of Africa.

The great apes are indeed physically similar to us, and they are indeed our closest living relatives. Particularly the African great apes: the familiar common chimpanzee (*Pan troglodytes*), found across Central and West Africa; the bonobo (*Pan paniscus*), a very similar and rare species with a distinctive black face and a head with long hair parted in the center; and the larger gorilla (*Gorilla gorilla*), rare in the Virunga mountains but more common in the lowlands of West Africa. None is plentiful in nature; all are restricted to patchy and often discontinuous ranges of land in a quite restricted part of the world. Their cousin, *Homo sapiens,* is by contrast exceedingly plentiful.

Too plentiful, many would argue.

We are slightly more distantly related to the red-haired orangutan of Borneo and Sumatra. Like the chimpanzees and gorillas, the Asian orangutan is also a "great ape," in contrast to the small, long-armed, and exceedingly graceful gibbons of southeast Asia, the "lesser apes." We're still similar enough to the orangutans, though, that the very name "orangutan" means "man of the woods" in Malay. The indigenous peoples didn't need science to tell them of that creature's general similarities to them.

The general estimate, which there is no good reason to doubt at present, is that about seven million years ago *Homo, Pan,* and *Gorilla* all comprised a single species. That species lived in Africa (which is, after all, where its descendants live), and probably resembled the chimpanzee. One group evolved a larger body size and ultimately became gorillas, and another group began to walk upright and ultimately became humans. And it is certainly possible that there were several other branches of that family tree that flourished, say, three million years ago, but whose remains have not yet been unearthed.

Obviously, a lot can happen in seven million years—a creature resembling a chimpanzee can evolve into a creature resembling a human being, for example. But it is also paradoxically little more than a flash in biological time. It all depends upon one's perspective. The average species of clam, for example, remains largely unchanged for as much as ten million years.

Fifteen million years ago, there were many diverse species of apes, thriving in the pristine and abundant woodlands of Africa, Asia, and Europe. The Miocene epoch, which lasted from about twenty-five to five million years ago, encompasses the florescence of the apes. They ranged in size from the diminutive *Micropithecus* to the aptly named *Gigantopithecus,* whose jaw dwarfs that of a modern gorilla. All looked diagnosably different from one another and from modern apes, but all would be recognizable as apes; none would be readily mistaken for a raccoon or a billy goat.

Modern apes thus constitute in the present day but a minuscule relic of the ecological space once occupied by their mighty group. And this shift from diverse and prolific to nearly extinct, from masters of the primate world to its tattered and pathetic remnants, has occurred within an evolutionary filament of the billions of years of life on earth, the single thread that subsumes the emergence and development of the human line from that ape radiation.

Seven million years ago, toward the end of the Miocene, the ancestors of living humans, gorillas and chimps went their separate evolutionary ways. Possibly the proto-chimpanzees were spread widely over Central Africa, the proto-gorillas occupied West Africa, where we find modern gorillas, and the proto-humans occupied East Africa, where we find the earliest humanlike fossils.

The human fossils are more copious than later ape fossils because, like modern apes, the ancestral apes lived generally in forested areas whose soil doesn't preserve fossils well. The human hallmark, bipedalism, took our ancestor to open savannas, where the complex geological processes of fossilization could more readily occur. The resulting abundance of proto-human fossils and near absence of later ape (chimp/gorilla) fossils makes it unlikely that the fossil record will tell us much about the actual historical divergence of humans and living apes from one another.

But we do know that the historical divergence was quite recent. In the mid-1960s, a collaboration between two scientists at Berkeley—a soft-spoken, slender, New-Zealand-born biochemist, Allan Wilson, and a swarthy, genial, and argumentative six-and-a-half-foot anthropologist, Vincent Sarich—demonstrated it in a classic series of papers.

Although by then the genetic code had been worked out, DNA could not yet be directly analyzed. Instead, one used a surrogate measure of genetic difference, the proteins. Proteins are a direct reflection of DNA, being the products of the DNA code, and consequently they can be used as reasonable first estimates of genetic similarity. Morris Goodman, a biochemist at Wayne State University, had recently demonstrated that the proteins of humans, chimpanzees, and gorillas were extraordinarily similar. Sarich and Wilson, however, wanted to take the research to another level by asking just how similar the proteins were, putting numbers on Goodman's crude observation.

Their technique involved injecting a rabbit with a human protein. As would the blood of most vertebrates, the rabbit's blood recognizes the human protein as a foreign substance, like a virus, and produces antibodies to attack it. The immune reaction is highly specific and highly efficient, and once it has "learned" to attack the human substance, the rabbit's blood will now also react against a similar protein from a baboon. But since the baboon protein is only similar to, but not identical with, the human protein, the rabbit's blood reaction is somewhat weaker.

How much weaker? By studying the extent of the reaction of rabbit blood immunized against human proteins to the proteins of, say, an orangutan, Sarich and Wilson could estimate the amount of difference between the human and orangutan proteins as simply the difference in the intensities of the chemical reactions in the rabbit blood.

On the other side of the country, an ambitious young British paleontologist at Yale named David Pilbeam was arguing that the jaw fragments of a fourteen-million-year-old fossil called *Ramapithecus* showed the dental hallmarks of humanity. If they indeed evinced humanity and they were that old, it followed that humans had diverged from the apes at least fourteen million years ago.

But Sarich and Wilson had come up with an interesting result. They found that rates of change in the protein varied very little from species to species. Humans and orangutans are fairly close relatives, from whom the baboon diverged much earlier. The ancestors of humans and orangutans were consequently the same species back when they diverged from the baboon. And Sarich and Wilson found that the intensity of reaction was generally the same if you studied the human-baboon reaction or the orangutan-baboon reaction.

There just didn't seem to be much in the way of feverish activity in the evolution of the proteins of humans, orangutans, or baboons. Rates of protein evolution simply did not seem to vary extensively. The amount of difference seemed sensitive only to the time since the species had been separated from one another.

Knowing that the rates of protein evolution were roughly constant, Sarich and Wilson calculated those rates and then applied them to the human-chimpanzee divergence. Knowing how rapidly change occurs in a stretch of evolutionary time, and how much protein change has accumulated since the divergence of human from chimpanzee, it was a straightforward matter to calculate how long humans and chimpanzees had been independent evolutionary lineages.

Their answer: about four million years.

The problem was that based on the *Ramapithecus* jaw fragments, Pilbeam was stridently proclaiming the independence of the human lineage over fourteen million years ago, while Sarich and Wilson were arguing just as forcefully that *Ramapithecus* couldn't be in the human line, because there had been no human line that early; fourteen million years ago, humans and modern apes had necessarily constituted the same evolutionary lineage. Even allowing for a bit of sloppiness in their calibration, Sarich famously wrote in 1970, "one no longer has the option of considering a fossil specimen older than about 8 million years as a hominid *no matter what it looks like*."

Looks, as we all know, can be deceiving, and *Ramapithecus* has since been shown not to have been a human ancestor. Details of its face show it to have been more closely related to the orangutan.

But it would be wrong to read this as a victory of genetics over anatomy. It was, rather, the a triumph of one carefully interpreted corpus of data over another body of data, interpreted more freely. In this case, the superior scholarship and science lay with the biochemists. Under slightly different circumstances, it could just as easily have been the other way around.

And often it is—surprisingly often, in fact. We routinely now hear, however, of the reassessment of phylogenies based on molecular data—that "no matter what it looks like," cockeyed genetic evidence shows that guinea pigs aren't really rodents, that dogs were domesticated ten times earlier than had been thought, that humans diverged from Neandertals five times earlier than had been thought, that rabbits are more closely related to people than to mice, that frogs are more closely related to fish than to lizards, and that the groups of modern mammals diverged from one another far earlier than everyone thinks.

Even Vince Sarich, the doyen of iconoclastic molecular evolutionary studies, takes a jaundiced view today from retirement. Small miscalculations can have huge effects, he notes.

> Then toss in a bit of molecular omnipotence, mix in a disdain for
> the paleontologists, and you've got me some 25 years ago—and
> other researchers today. How did I escape? Well, mostly by reading,
> listening, and thinking. Having been down the very seductive
> "molecules are everything; fossils are nothing" road, and through
> the lengthy and painful process of weaning myself off it, it hurts

me to see that the existence of the lesson is not even acknowledged, never mind the lesson itself being learned.

Perhaps the most overexposed factoid in modern science is our genetic similarity to those African apes, the chimpanzees and gorillas. It bears the precision of modern technology; it carries the air of philosophical relevance. It reinforces what we already suspect, that genetics reveals deep truths about the human condition; that we are only half a step from the beast in our nature. Chimpanzees can be brutal, humans can be brutal; it's in our essence, it's in our genes.

But how do we know just how genetically similar we are to them? What is that estimate based on? What real significance does it have for our conceptions of ourselves in the modern world, and the role of genetic knowledge in shaping those conceptions? This is where genetics and anthropology converge, the gray zone of molecular anthropology, which forces us not just to look at the genetic data but to question both the cultural assumptions we bring to those data and their relevance for thinking about the modern world and interpreting our place within it.

SCIENCE CONFRONTS THE CHIMPANZEE

The fact that we are biologically similar to the apes was known long before there were geneticists. For eighteenth-century scholars, apes had roughly the same status as Bigfoot does today: they lived in remote areas and were seen only by untrained observers. Consequently, reports about them differed widely in quality and reliability.

One famous and influential story was that of a sixteenth-century sailor named Andrew Battell, who survived a shipwreck and lived in West Africa for several years before returning to Europe. His story was related in a popular collection of travel narratives published by Samuel Purchas in the early 1600s. Purchas wrote of "two kinds of Monsters, which are common in these Woods, and very dangerous," known in the native tongue as the "Pongo" and the "Engeco." Battell had nothing to say about the Engeco, but the Pongo's "highth was like a mans, but their bignesse twice as great."

Could he have been describing a gorilla?

> This Pongo is in all proportion like a man, but that he is more like a Giant in stature, then a man: for he is very tall, and hath a mans face, hollow eyed, with long haire upon his browes. His face and eares are without haire, and his hands also. His bodie is full of haire, but not very thicke, and it is of a dunnish colour. He differeth not from a man, but in his legs, for they have no calfe. Hee goeth alwaies upon his legs, and carrieth his hands clasped on the nape of his necke, when he goeth upon the ground. They sleepe in the trees, and build shelters for the raine. They feed upon Fruit that they find in the Woods, and upon Nuts, for they eate no kind of flesh. They cannot speake, and have no understanding more then a beast. The People of the Countrie, when they travaile in the Woods, make fires where they sleepe in the night; and in the morning, when they are gone, the Pongoes will come and sit about the fire, til it goeth out: for they have no understanding to lay the wood together. They goe many together, and kill many Negroes that travaile in the Woods. Many times they fall upon the Elephants, which come to feed where they be, and so beate them with their clubbed fists, and pieces of wood, that they will runne roaring away from them. Those Pongoes are never taken alive, because they are so strong, that ten men cannot hold one of them: but yet they take many of their young ones with poisoned Arrowes. The young Pongo hangeth on his mothers bellie, with his hands fast clasped about her: so that, when the Countrie people kill any of the femals, they take the young one, which hangeth fast upon his mother. When they die among themselves, they cover the dead with great heapes of boughs and wood, which is commonly found in the Forrests.

Perhaps that passage records a genuine encounter with gorillas. On the other hand, this was an era of exploration, an era of great discoveries of lands, peoples, and animals. We now know that gorillas sleep in trees, eat no meat, and do not speak. However, unless gorillas have changed a lot in the past four hundred years, we also know that they are not bipedal, and practice no funerary rites. Was this Pongo, then, more like the well-known baboon or the well-known centaur?

The fact is, however, that the narrator identifies the Pongo and the Engeco as "Monsters" and distinguishes them from "Baboones, Monkies, Apes, and Parrots"—rendering the identification of the creature even more ambiguous and attesting at very least to the unfamiliar manner in which scholars of that era thought about the animal kingdom.

These monsters were situated on the boundary between personhood and animalhood. Consequently, they were immensely interesting. That boundary is the domain of powerful mythological motifs in all cultures, for the distinction between person and animal allows us to situate ourselves in the natural order, to make some sense of our place in it. And the mythology is just as powerful in the scientific culture, as the scientific literature will easily attest: these creatures are both "us" and "not-us," and *we need to know what they really are.*

When European sailors traveled to Africa and Asia, they often tried to retrieve these creatures, but unfortunately the journey was long and hard, and a baby chimpanzee could probably not be expected to thrive on a seafaring diet of biscuits and salt pork. An ape of some sort was described briefly by a Dutch anatomist named Nicolaas Tulp in 1641, but the terse description of the animal is so confusing that it is hard to know whether Tulp had been brought a chimpanzee or an orangutan. Tulp's "satyr" (fig. 1) looks more like an orangutan than like anything else, and he tells us that "it is called by the Indians orang-outang," but he also says his specimen didn't come from the Indies, but from Angola, and it had black hair.

"The face," Tulp relates, "counterfeits man: but the nostrils are flat and bent inwards, like a wrinkled and toothless old woman." But the ears, breast, abdomen and limbs were so like ours that he considered them as similar as two eggs in a basket. The satyr walked upright and had prodigious strength; yet it drank with great delicacy and "afterward wiped away the moisture on its lips not less suitably and less

Figure 1. Tulp's "satyr." From Nicolaas Tulp, *Observationum medicarum libri tres* . . . (Amsterdam: Apud Ludovicum Elzevirium, 1641).

delicately than thou wouldst see it in the court of princes." It even slept like a person.

Yet the satyr had a dark side. In Borneo,

> they not only make attacks on armed men, but upon women and girls. Meanwhile the desire for the latter burns so ardently that not seldom they ravish them when captured. In fact they are so greatly inclined to venery (even among themselves, as was common with the licentious Satyrs of the ancients) that they are at all times

wanton and lustful: so that the Indian women therefore avoid
the woods and forests . . . where these shameless animals roam.

Clearly, Tulp is conflating classic mythology, actual observations, and the symbolic criteria popularly used to demarcate the human condition (we are prim and sexually modest; they are immoral and licentious—a recurrent theme when dealing with other species and other races). And that is precisely the problem—the apes, by virtue of straddling a symbolic boundary, are highly subject to the projections of the scientist from the very outset of modern science.

The first successful description in the scientific literature of a great ape appeared in 1699 by the leading anatomist in England, Edward Tyson. Picked up in Angola and brought to London by ship, the chimpanzee succumbed shortly thereafter to complications stemming from a jaw infection it sustained from a fall on board. However, the anatomist saw the animal alive, dissected it afterward, and described and illustrated his work in such detail that it is impossible to mistake his subject for anything but a chimpanzee (fig. 2).

He called it a "Pygmie," and was struck by the fact that when he tallied up the resemblances of the Pygmie to a monkey, there were thirty-four; but when he tallied up to the resemblances to a human, he counted forty-eight. It was clearly like a human: "It would lie in a bed, place his head on the pillow, and pull the cloathes over him, as a man would do; but was so careless and so very a brute, as to do all Nature's Occasions there."

But it was clearly not a human: "Now when I observed it to go upon all four, . . . it did not place the palms of the hand flat to the ground, but went upon its knuckles, . . . which seemed to me so un usual a way of walking, as I have not observed the like before in any animal."

Tyson resolved the paradox by emphasizing the physical similarities and the mental or spiritual discontinuities. The "Pygmie" partook of both essences, the physical human and the mental nonhuman, and thus formed a link between the material world of animals and the spiritual world of people.

But the sterile analysis of the scientist-philosopher did not answer

Figure 2. Tyson's "orang-outang," or "pygmie."
From Edward Tyson, *Orang-outang, sive homo sylvestris, or, The anatomy of a pygmie* (London: Thomas Bennett & Daniel Brown, 1699).

the question of what the creature was like when it was alive—how it lived, how it moved, what it ate. For this, Europeans were still obliged to rely on the accounts of sailors, travelers, or merchants. A contemporary of Tyson's, Willem Bosman, wrote a popular Dutch account of his travels in West Africa, translated into English in 1705. He relates that there are "so many various species" of apes that he can't possibly describe them all, but begins with apes he calls "Smitten," which are of a pale mouse color, and grow to a wonderful size. I have myself seen one of five foot long, and not much less than a man; they are very mischievous and bold. It seems incredible what an English merchant here affirmed to me for truth; that behind the English fort at Wimba (where there is a terrible number of these apes; that are so bold, that they will attack a man, as he related,) amongst others they fell upon two of their Company's slaves, which the apes had overpowered, and would have poked out their eyes, if they had not been timely rescued by some Negroes; for they, to complete their design, had gotten some sticks ready.

You, as well as myself, are at liberty what credit to give to this story. But indeed, these are a terrible pernicious sort of brutes, which seem to be made only for mischief.

Some of the Negroes believe, as an undoubted truth, that these apes can speak, but will not, that they may not be set to work; which they do not very well love: this is their opinion of them.

The *Speaker,* an English merchant ship, docked in London in 1738. She had brought from Angola "an animal of remarkably and terribly hideous countenance . . . called by the name Chimpanzee." This had the distinction of being the first ape widely displayed in Europe, and it impressed a number of leading scholars with its ability to "ape" the human form and action. Again, the animal, a youngster, was made the subject of a scientific report, which noted its habits of

walking with an erect body, with a few parts of the body being covered with hair, the rest physically strong and muscular. It seeks food from its own excrement; but it also loves the drink of tea, which it drinks from a small container in the manner of humans. Moreover it imitates the sleep of these humans and is not altogether lacking in intelligence, even expressing a human chattiness in its own voice. The males, when they have come to the age of adulthood, seek out human females for illicit sexual intercourse, and provoke men, even armed ones, to fighting.

Biological systematics—how we formally organize and partition nature—matured through the work of a Swedish botanist and physician, Carl Linnaeus. Biologists as far back as Aristotle had classified animals; but Linnaeus succeeded in imposing regularity and rigor on the process.

The father of scientific classification was an eighteenth-century scholar of humble origins. Born in 1707, the son of a rural clergyman, Linnaeus demonstrated a facility for botany that impressed the professors at Sweden's most distinguished university, Uppsala. Arriving at the university in poverty, he resoled his classmates' cast-off shoes to wear himself, but he soon managed to earn his keep as a tutor. He obtained a medical degree in Holland in 1735; and that very year he published the first edition of his *Systema naturae* (System of Nature), using the sexual anatomy of plants as a key to classifying them. In the next three years, he published no fewer than eight books. He then returned to Sweden to marry.

Linnaeus hung out his shingle in Stockholm to practice medicine, but his reputation as a scholar was such that he was appointed to the medical faculty of Uppsala in 1741. Within a year, he had contrived to take charge of botany. An inspiring teacher and prodigious writer, Linnaeus supervised over 180 doctoral theses during his academic career and took an active role in writing them as well. His students and colleagues sent him plant specimens from all over the world, and Linnaeus fitted them all into the system of nature. It was said that God created, but Linnaeus arranged.

Nature, according to Linnaeus, in keeping with the new scientific thought of the eighteenth century, could be expressed in simple formulas. What Newton had achieved mathematically with motion, Linnaeus could do anatomically with species. Prickly, insecure, obsessed with his place in history, and unable to accept criticism, Linnaeus had not quite as extreme a personality as the legendary Newton, but he was known as someone you had to handle gingerly.

Linnaeus's *Systema naturae,* an authoritative guide to the arrangement of plants, animals, and minerals as ordained by God and discerned by the author, went through twelve editions, and it is from the tenth (1758) that modern biological classification officially dates. He was eventually ennobled, retroactively to 1757, under the name Carl von Linné, and died in 1778.

And what sense did Linnaeus make of the relationship between people and apes? The father of zoological classification was so confused that he simply divided reports about them into two sets, the more anthropomorphic and less anthropomorphic. He designated the former as a second species of humans, *Homo troglodytes* (which he also called *Homo nocturnus*—nocturnal, or cave-dwelling, man), and put the latter in another primate genus, *Simia satyrus.* Not only were the apes paradoxically very much like us and very much not like us simultaneously, but they were now formally both human and non-human at the same time.

Subsequent generations of biologists quickly corrected Linnaeus's error. As early as 1771, an English zoologist named Thomas Pennant referred to "Linnaeus's Homo nocturnus, an animal . . . unnecessarily separated from his Simia satyrus." In 1775, the German biologist Johann Friedrich Blumenbach noted that Linnaeus had made a "great mistake" in the *Systema naturae,* in "that the attributes of apes are there mixed up with those of men."

Biologists since the middle of the eighteenth century have sought to highlight one or the other side of this paradox. Thus, we follow Linnaeus today in classifying humans as "just" another species of primate, specifically "just" another species of apelike creature. The official version follows the paleontologist George Gaylord Simpson's 1945 monograph on classifying mammals, placing us in the superfamily Hominoidea, along with the lesser apes (gibbons) and great apes (chimpanzees, gorillas, and orangutans). The special similarities of structure that unite us with these creatures among other primates generally are their teeth (since most of the vertebrate fossil record is teeth, these are important diagnostic features), the lack of a tail, the position and movement capabilities of the shoulder, and the structure of the trunk.

We might alternatively choose to emphasize the "otherness" of humans by dividing the primate group fundamentally into "Quadrumana" and "Bimana," or two-handed and four-handed, as was popular among Linnaeus's close intellectual descendants. Primates, being generally arboreal creatures, are different from other mammals in the structure of their hands and feet, which are grasping appendages and lack the claws that other arboreal mammals use. Humans, of course, retain this heritage in their hands, but not in their feet; we are not gracefully four-handed but pitiably two-handed. This classification would acknowledge the unique aspects of human feet, which are specialized unlike those of any other primate and have lost the ability to grasp, except in a very rudimentary manner. It wouldn't deny the common ancestry of humans and apes, but would merely highlight the divergence of humans from their ape ancestry.

We could even separate humans from other multicellular life altogether, placing humans in the subkingdom Psychozoa—"mental life"—as Julian Huxley proposed half a century ago. A species that lives by its wits, that is to say, that relies entirely on the technological products of its societies for individual survival, is quite different from other life on earth. Perhaps indeed that might be worth acknowledging zoologically.

Humans are marked by a large number of physical, ecological, mental, and social distinctions from other life. Not improvements, of course—we have no objective way to evaluate "improvement" in the natural world—but merely differences. Other primate species spend time on two legs, and other vertebrate species are bipedal (birds and kangaroos come readily to mind), but not in the same manner as humans. Other species communicate, but not via the absurdly arbitrary and symbolic media we call language. Other species modify natural objects and use them to aid in feeding (as noted in Jane Goodall's classic observations of chimpanzees stripping twigs and using them to "fish" for termites), but none rely on their technology to survive as humans do. And in no other species does technology take on an evolutionary trajectory of its own, a result of the social cycle of invention, adoption, spread, and modification. Other species appear to grieve, but none weep as humans do—and certainly not over imaginary events like those in *Les Misérables* or *Love Story*.

What does genetics have to say about all this?

Nothing.

Sameness/otherness is a philosophical paradox that is resolved by argument, not by data. Genetic data tell us precisely what we already knew, that humans are both very similar to and different from the great apes.

But genetics is able to place a number on that similarity. It is not uncommon to encounter the statement that we are something like 99.44% genetically identical to chimpanzees. (Actually that number is the fabled degree of purity of Ivory Soap, but it serves the purpose.) You can literally count the number of base differences among the same regions of DNA in humans and chimpanzees and gorillas and add them up. Or you can do the same thing to the products of the DNA, protein structures. All such comparisons invariably yield the result that humans and chimpanzees (and gorillas) are extraordinarily similar. But that was known to Linnaeus without the aid of molecular genetics.

So what's new?

Just the number.

GENETICAL AND ANATOMICAL CONTRASTS

If you compare a human and a chimpanzee, it is easy to see that structurally they are remarkably similar. Every bone of the chimpanzee body corresponds almost perfectly to a bone in a human body—but differs ever so slightly and diagnostically, in ways generally related to the human habit of walking upright. And if not related to our bipedal habit, any detectable difference is very likely related to either of two other human physical specializations, our reduced front teeth and our enlarged brains.

The problem is simply that it's difficult to say just how similar a particular chimpanzee body part and its human counterpart are, percentagewise. Where a human skull encases about 1400 cubic centimeters of brain, a chimp is lucky to have a third of that. Is that 67% different? Where a chimp's elbow has large bumps to accommodate the muscles that bend it, a human has smaller bumps; where a human's thighbone has the lower part, corresponding to the knee, pointing in the same direction as the upper part, corresponding to the hip, a chimp's thighbone is twisted, so its knee points outward rather than forward. Just how different is that?

A percentage, after all, is a scalar, one-dimensional measure, while a body part is a three-dimensional entity. Indeed, four-dimensional, if you consider the developmental aspects of growth—as a person matures, they change in form, they don't merely expand.

How similar is a large sphere to a small sphere? Is a large sphere more similar to a small sphere or to a large cube?

Obviously, it depends on which criterion you decide is most important—size or shape. Where there are two different and incomparable kinds of similarities being observed here, an evaluation along a single scale is impossible; in fact, it's absurd. If you were to compare human and chimpanzee structure in the context of a three-way comparison to, say, a snail, it would be patently obvious that humans and chimps are over 99% identical in practically every way you could compare all three animals.

The human and chimp have bones, they lack a shell, they have limbs, they don't leave a trail of slime as they move; every nerve, every sinew, every organ is almost the same in a chimp and human and very different, if present at all, in the snail.

Exactly how similar they are, of course, remains elusive. Even assuming it to be true, it would be rather old-fashioned and premodern to say, "Humans and chimpanzees are really, really, really similar," even though it is true.

What genetics offers is the opportunity to place a hard number on two-way comparisons, by virtue of the fact that the genetic instructions (and their primary products) are composed of long chains of subunits, differences in which can be tabulated and numerically manipulated. Thus, I can look at a short region around one of the genes for hemoglobin, as my colleagues and I did in 1986, and find a DNA sequence that reads GCTGGAGCCTCGGTGGCCAT in a baboon, and GCTGGAGACTCGGTGGCCAT in an orangutan. One difference in twenty possibilities; the very linearity of DNA sequences makes them easy to compare. And with a much longer DNA sequence, you would expect to get a more accurate estimate. In this case, for example, we found a difference of about 5% in this small region between a baboon and an orangutan, and a more general level of difference of about 8%.

But the comparison can also be misleading in two important ways.

First, such comparisons of DNA sequence ignore qualitative differences, those of kind rather than amount. To take the smallest case, consider a different sequence of twenty DNA bases from the same region: CCTTGGGCCTCCCGCCAGGC in the baboon and CCTTGGGCTCCCGCCAGGCC in the orangutan. If you look at them in parallel rows you find them to be different.

> ... CCTTGGGC*CT*CC*CG*C*CA*G*G*C ...
> ... CCTTGGGC*TC*CC*GC*C*AG*G*CC ...

But if you look more carefully, you might observe that the general gestalt of the sequences is roughly the same. If there is one "C" too many in the middle of the top sequence, or one two few in the bottom, the match becomes far more complete. (Since DNA sequences are immensely long, the presence of extra bases at the beginning or end of an arbitrary sequence is trivial. The ones in the middle of sequences are interesting.) If we look at it again, inserting a small gap for one base too many or too few, we see the similarity that was previously hidden from view.

> ... CCTTGGGCCTCCCGCCAGGC ...
> ... CCTTGGGC TCCCGCCAGGCC ...

So we may infer an insertion in one lineage or a deletion in the other, in order to make this sequence look maximally similar. But how can we know for sure? This is a very different kind of inference from the DNA base substitutions we were tabulating a few paragraphs ago. This involved ignoring the actual census of differences in favor of a "gestalt" similarity, and then retabulating the differences—a highly subjective procedure, although probably right.

So we've overridden the observation of seven differences in twenty with the inference of one difference, of a different sort, in twenty. Again, this is not to say that it's illegitimate (the example is actually taken from a paper I co-authored), but the question is, what does it do to our number, our precise estimate of the degree of genetic difference? First, it inserts an element of subjectivity masked by the number itself; and second, it sums together DNA base substitutions and DNA base deletions, as if they were biochemically identical and quantitatively equivalent. In fact they are neither; this is a classic "apples and oranges" case.

And that was an easy one.

In fact, the molecular apparatus has complex ways of generating insertions and deletions in DNA, which we are only beginning to understand. From a more poorly understood genetic region (the genes for ribosomal RNA), here is a DNA sequence from human and orangutan, aligned three different ways:

```
Human       CCTCCGCCGCGCCG      CTCCGC GCCGCCGGGCA              CGGCC           CCGC
Orangutan   CC                  GTCGCCTCCGCCACGCCGCGCCACCGGGCCGGGCCGGCCCGGCCCGCCCGC

Human             CCTCCGCCGCGCCGCT        CCGCGCCGCCGGGCACGGCCCCGC
Orangutan   CCGTCGCCTCCGCCACGCCGCGCCACCGGGCCGGGCCGGCCCGGCCCGCCCGC

Human       CCTCCGCCGCGCCG      CTCCGCGCCGCCGGG CAC  GGCC              CCGC
Orangutan               CCGTCGCCTCCGCCACGCCGCGCCACCGGGCCGGGCCGGCCCGGCCCGCCCCGC
```

These are precisely the same sequences, reconstructed and presented three different ways. Tabulate the differences. The top one invokes five gaps and six base substitutions; the middle has only two gaps but nine base substitutions. And the bottom one has five gaps and only three base substitutions. The three pairs of sequences differ in the assumptions about which base in one species corresponds to which base in the other. While we might, by Occam's Razor, choose the alignment that invokes the fewest inferred hypothetical evolutionary events, we still have to decide whether a gap "equals" a substitution. Does the bottom one win because it has a total of only eight differences? Or might the middle one win because a gap should be considered rare and thereby "worth," say, five base substitutions?

The problem is that we cannot know which is "right," and the one we choose will contain implicit information about what evolutionary events have occurred, which will in turn affect the amount of similarity we tally. How similar is this stretch of DNA between human and orangutan? There may be seven differences or eleven differences, depending upon how we decide the bases correspond to each other across the species—and that is, of course, assuming that a one-base gap is also equivalent to a five-base gap and to a base substitution.

In a more general sense, however, the problem of taking quantitative estimates of difference between entities that differ in quality is prevalent throughout the genetic comparison of human and ape. The comparison of DNA sequences presupposes that there are corresponding, homologous sequences in both species, which of course there must be if such a comparison is actually being undertaken. But other measurements have shown that a chimpanzee cell has 10% more DNA than a human cell. (This doesn't mean anything functionally, since most DNA is functionless.) But how do you work that information into the comparison, or into the 99.44% similarity?

In the example just given, you may notice that our problem started with the assumption that forty DNA bases of human sequences were homologous to fifty-four DNA bases of orangutan sequences. A simple estimate of similarity and difference must necessarily be confounded by variation in the size of the entities being compared.

If we compare the genes for alpha-hemoglobin (half of the molecule that transports oxygen and carbon dioxide in our blood) on chromosome 16 between human and chimpanzee, we find a near identity of the base sequences. But we also find that, with rare exceptions, humans have two copies of the alpha-hemoglobin gene, aligned in tandem, whereas chimpanzees have three. Or if we look at the genes that code for the Rh blood group, again the nucleotides of the genes match up almost perfectly between the two species, but humans have two such genes, and chimpanzees have three. How can you make simple numerical sense of that?

The second misleading area of DNA sequence comparisons entails a consideration of the other end of the scale. The structure of DNA, the famous "double helix," is built up of four simple subunits. Each of our cells has a length of DNA encompassing approximately 3.2 billion of these subunits, but there are still only four of them: adenine, guanine, cytosine, and thymine, or A, G, C, T. This creates a statistical oddity. Since genetic information is composed of DNA sequences, and there are only four elements to each DNA sequence, it follows that two DNA sequences can differ, on the average, by no more than 75%. Certainly a very small stretch of DNA might be 0% similar to another very small stretch (AAAAT matches GGCCG nowhere, after all), but on the average two random stretches of DNA will be statistically obliged to match at one out of four places.

In other words, two stretches of DNA generated completely at random, completely independently of one another, would not be 0% similar but, rather, would be 25% similar.

But what would constitute a comparison of two DNA segments that emerged completely independently of one another? When we compare DNA sequences, of course, we are comparing corresponding DNA sequences, such as the gene that codes for the electron-transport protein cytochrome c. Such correspondence, or homology, in biology is a reflection of common descent. A human and a chimpanzee have similar genes for cytochrome c because they are descended from a common ancestor that had a gene for cytochrome c similar to those of both.

In fact, so do humans and fruitflies. Their common ancestor is more remote in time than that of human and chimpanzee, but the fly has the molecule cytochrome c, it does the same job as the human and chimpanzee versions, and the genes can be compared precisely because humans and flies share a common ancestry. Their DNA sequences did not emerge independently of one another but are products of the divergent histories of lineages that became separated some hundreds of millions of years ago.

In fact, so do humans and daffodils. Their DNA sequences are not independent of one another either. Any such independent DNA sequences would have to result from an independent origin of DNA-based life; and then we would still expect them to be 25% identical, by virtue of the way in which DNA is constructed.

Thus, if you compare the DNA of a human and a daffodil, the 25% mark is actually the zero mark, and since humans and daffodils do share a common ancestry, you would expect them to be generally more similar than 25%. Say, about 35%.

In the context of a 35% similarity to a daffodil, the 99.44% similarity of the DNA of human to chimp doesn't seem so remarkable. After all, humans are obviously a heck of a lot more similar to chimpanzees than to daffodils.

More than that, to say that humans are over one-third daffodil is more ludicrous than profound. There are hardly any comparisons you can make to a daffodil in which humans are 33% similar. DNA comparisons thus overestimate similarity at the low end of the scale (because 25% is actually the zero-mark of a DNA comparison) and underestimate comparisons at the high end. At least snails, in the earlier anatomical comparison, move around and eat; they can't photosynthesize. So from the standpoint of a daffodil, humans and chimpanzees aren't even 99.44% identical, they're 100% identical. The only difference between them is that the chimpanzee would probably be the one *eating* the daffodil.

NINETY-NINE AND FORTY-FOUR ONE-HUNDREDTHS PERCENT PURE HUMAN

That particular Ivory Soap number is itself interesting. In a recent book on the great apes, that specific similarity between human and chimpanzee is invoked by several different authors. On p. 12 we learn that "our DNA differs from theirs by only just over 1 per cent"; on p. 39 we are "within 1 per cent"; on p. 95 that difference is put at "1.6 per cent"; and on p. 220 we are told that they "overlap more than 98 per cent of their genes with us."

Whether the actual number is less than 1% or 2% is obviously trivial in the great scheme of things. Our genes are very, very similar. But somehow "very, very similar" doesn't sound quite scientific enough. The invocation of a number with a decimal point, on the other hand, certainly does sound scientific. It implies that there is some officially calculated, scientifically sanctioned degree of similarity between the DNA of human and chimpanzee, when in fact there are simply a variety of crudely measured, but generally concordant, studies.

The problem is that in being told about these data without a context in which to interpret them, we are left to our own cultural devices. Here, we are generally expected to infer that genetic comparisons reflect deep biological structure, and that 98% is an overwhelming amount of similarity. Thus "the DNA of a human is 98% identical to the DNA of a chimpanzee" becomes casually interpreted as "deep down inside, humans are overwhelmingly chimpanzee. Like 98% chimpanzee."

The important question that lurks behind this trivial disagreement is, What constitutes a scientific statement about genetic similarity? The estimates of genetic difference between human and chimpanzee given in that book are precise, but not accurate. In other words, they have a decimal point and are written out to a tenth of a percent, but the numbers given are wrong—you don't have to be Aristotle to see that it can't be both 1.6% and within 1% at the same time. The fact that either or both are inaccurate doesn't seem to make them less scientific.

The cachet of science resides here in both the DNA and the tenths-of-percentiles. DNA is now the stuff of bad direct-to-video movie plots and a sweet-smelling cologne from Bijan of Beverly Hills, which even comes in a helical bottle. (The fact that the helical bottle is composed of three strands, rather than two, earned the company the IgNobel prize for chemistry in 1996, presented at Harvard by the *Annals of Improbable Research.* The spectacular, hilarious award ceremony has achieved cult status in the academic community.) And the allure of tenths-of-percentiles is surely what Benjamin Disraeli had in mind when he grouped statistics along with lies and damn lies.

Geneticists have occasionally fallen headlong into this trap of taking precision in lieu of accuracy. For example, in the 1920s, it was not known how many chromosomes a human being had in each cell. Most counts were in the twenties (by American scientists), some were as high as the forties (by European scientists), and some geneticists reconciled the discrepancy by arguing that since the Americans were studying the chromosomes of black people and the Europeans were studying those of white people, perhaps whites had more chromosomes than blacks. In 1927, however, the leading cell biologist of the era pronounced definitively on the human chromosome number, forty-eight. The count of forty-eight entered the textbooks and remained the official human chromosome number for three decades; indeed, the count was replicated several times. That is how these scientists knew they were doing their experiments right; they identified all forty-eight human chromosomes.

In fact, these scientists were attempting essentially to count the strands in a plate of linguini, which is daunting, at best. The truth, the accurate inference, was always that humans had a large number of chromosomes, probably in the high forties. But that wasn't what they said. With the development of new techniques in cell biology, it emerged in 1956 that humans really only have forty-six chromosomes.

Whoops!

The scientists were accepting a count that was precise over a count that was true. Thus, the scientific statement of thirty years' standing was precise, authoritative, and wrong.

And all the geneticists were doing was counting.

And so it is with the 99.44% genetic identity to a chimpanzee. We don't know precisely how similar the DNAs of chimps and humans are, except that they are very similar, as are their bodies. Genetics appropriates that discovery as a triumph because it can place a number on it, but the number is rather unreliable as such. And whatever the number is, it shouldn't be any more impressive than the anatomical similarity; all we need to do is to put that old-fashioned comparison into a zoological context.

The paradox is not that we are so genetically similar to the chimpanzee; the paradox is why we now find the genetic similarity to be so much more striking than the anatomical similarity. Scholars of the eighteenth century were overwhelmed by the similarities between humans and chimpanzees. Chimpanzees were as novel then as DNA is now; and the apparent contrast between our bodies and our genes is simply an artifact of having two centuries' familiarity with chimpanzees and scarcely two decades' familiarity with DNA sequences.

Two

THE APE IN YOU

WHERE DOES A NUMBER LIKE 99.44%, ostensibly representing our basic similarity to another species, come from? Three sorts of data have produced these tabulations. The first is a comparison of protein structure from 1975, by Berkeley geneticists Mary-Claire King and Allan Wilson. Proteins are long strands of elementary subunits, which reflect gene structure and can be compared. Examining the differences between forty-four known proteins of human and chimpanzee, King and Wilson found they were 99.3% identical. (Amazingly, only 0.14% away from the legendary purity of Ivory Soap!)

We know that about only 1% of the total DNA of a cell is actually expressed as proteins, and that protein-coding regions are the most slowly changing parts of the DNA. The reason is that mutations or changes to functional DNA are more likely to do systemic damage to the organism than mutations to nonfunctional DNA. Consequently, they are most unlikely to be perpetuated in the species, because their bearers fail to thrive and reproduce as efficiently as the other creatures who lack the damaging mutation. Mutations to non-functional DNA occur at the same rate, but because they are not expressed, they can do their bearers no harm. You would expect that DNA to differ more widely across species. Consequently, the King-

Wilson estimate of 99.3% is an overestimate of the similarity of human and chimpanzee DNA, being derived from a skewed sample of the DNA, reflecting only protein-coding regions.

The second class of data was made available in the mid-1980s with the development of direct DNA sequencing technology. DNA is a long linear molecule, again composed of simple subunits that can be compared. Bits and pieces of it have been compared between human and chimpanzee, amounting to less than 100,000 DNA bases in length. But there are 3.2 billion bases in the human genome, so obviously we have here a very minuscule proportion, and again there is a bias toward regions that contain functional units—genes—which tend to be where geneticists look for DNA to sequence.

The most comprehensive comparison we have is actually infinitesimal in scope, about 40,000 bases of the region of the hemoglobin genes on chromosome 11. And we find human and chimpanzee, base for base, to be about 1.9% different. The difference between the 0.7% estimate and the 1.9% estimate is a consequence of the fact that this DNA comparison includes much more nongenic DNA than genic DNA (which is the only class that would be translated into protein differences). Indeed, the DNA sequences of genes included in that region are virtually identical between the two species.

But again, that is a high estimate, because it focuses on a region known to contain genes and hence is more conservative than the overall DNA would be expected to be. Genes are rare in the genome. Not only that, but we know that the evolution of human hemoglobin has been sensitive to specific environmental problems—such as malaria—so it may have its own evolutionary idiosyncrasies.

Another piece of DNA that has been well studied is the mitochondrial DNA, or mtDNA. Where 3.2 billion nucleotides comprise the twenty-three chromosomes of a human nuclear genome (the DNA of a human gamete, half that of an ordinary cell), there is a tiny fraction located outside the nucleus. The mitochondrion, a subcellular organelle universally known in biology textbooks as "the powerhouse of the cell," generates metabolic energy for the physiological processes of life. It also has 16,500 nucleotides of DNA, which code for some of the molecules used by the mitochondrion. This mitochondrial DNA is much more famous for two reasons—it has suggested a primordial symbiotic relationship between free-living proto-mitochondria and free-living proto-cells at the dawn of life; and it is also the source of revolutionary inferences about modern human origins in a "mitochondrial Eve" (see chapter 4).

Less spectacularly, this DNA has also been compared across the great apes.

The mitochondrial DNA, however, is not 1% different among humans, chimpanzees, and gorillas. It is about 10% different. Why? Because mtDNA mutates at a much higher rate than nuclear DNA. The mutations have little or no effect on the life of the organism, and so the differences simply accumulate. Two organisms will therefore be far more similar in their nuclear DNA than in their mtDNA. For example, the "mitochondrial Eve" work was based on tabulating the differences detectable among human beings (0.2% difference), whose nuclear DNAs are so similar as to preclude that kind of study generally.

The point is an important one: different bits of DNA evolve at their own rates, and therefore in the six or seven million years of evolutionary time separating humans, chimps, and gorillas, some bits are 10% different, some are not comparable because they are different in kind rather than amount, and most are less than 3% different.

Another technique in the 1980s permitted a much broader estimate of DNA difference—a technique called DNA hybridization. Here, a pair of biologists from Yale adopted a technique that was recognized a decade earlier to be effective, if crude. They began by analytically dividing the DNA into two halves: repetitive DNA (redundant sequences of DNA, inferred to be meaningless) and unique-sequence DNA. Presumably, the latter contained the genes, so they discarded the former. Then, since DNA is a two-stranded molecule, the classic double helix, they separated the strands of human DNA (by heating it) and fooled the strands into rebonding with strands of chimpanzee DNA. Human DNA is easy to fool—if you simply give it a thousand times more chimpanzee DNA, it won't be able to find its perfect human match and will bond instead to whatever's around. The bonding to another species' DNA must be imperfect, however, since mutations have accumulated to distinguish them genetically as species.

They then heated the hybrid DNA molecules. It takes less heat to make these strands come apart, because they are imperfectly bonded—there are fewer molecular bonds holding them together. The difference between the temperature at which DNA strands from the same species come apart and those at which hybrid DNA strands come apart is thus a gross measure of how many mutations have occurred during the divergent evolutionary histories of those species.

The Yale biologists made two major claims. The first was that hybrid molecules of human and chimpanzee DNA dissociated from one another at a temperature 1.8 degrees lower than did pure human DNA. Assuming (conveniently) that 1 degree equals 1% genetic difference, they concluded that humans and chimpanzees are 1.8% genetically different. But there were two difficulties with this claim. First, others have calculated the conversion ratio at 1 degree equals 1.7% difference, which would make humans and chimps over 3% different, rather than 1.8%. And, second, the experiment only examined half of the total DNA in the first place. So it actually boils down to a demonstration that half the DNA of humans is either 98.2% or 97% identical to that of chimpanzees.

The Yale biologists' second claim was actually far more controversial. Not only were humans, chimpanzees, and gorillas genetically very, very similar (as was already well known), they asserted with great fanfare, but they could detect that humans and chimpanzees were actually more genetically similar to each other than either was to a gorilla. If true, this would be very significant.

They said that they had done the experiments over and over again and had always gotten values between 1.2 and 2.3 degrees for the human-chimp comparison and 1.6 to 2.8 for the chimp-gorilla comparison. Thus the genomes of human and chimps were only about 1.8% different (plus or minus a little), while each of them differed from the genome of a gorilla by about 2.3%. Unfortunately, however, when other scientists got to examine just one-eighth of these data, they found values for human-chimp ranging from -0.2 to 2.7 degrees—far beyond what the authors themselves had reported; and neither calculated "average" of 1.8 nor 2.3 was reliable—both sets of comparisons were simply about 2%. The numbers, alas, had been subjected to a battery of what were euphemistically called "correction factors" prior to publication, a fact conceded by the authors in a subsequent paper. They even admitted that had it not been for their unreported data alterations, they would have found humans, chimps, and gorillas to be equidistantly related.

That, however, would have been a non-story—failure to resolve the three-way split among the species with high-tech genetics and to establish the actual Ivory Soap value of the relationship. So the Yale experimenters ended up converting it into a story, the resolution of the genetic relationships of humans, chimpanzees, and gorillas, and an exact 98.4% genetic identity of humans with chimpanzees. If it had been true, it would indeed have been very interesting.

But it was simply another crude estimate of the genetic similarity of humans to apes—about 98%—unfortunately misrepresented as being more precise than it really was.

Yet another kind of comparison involves examining the structure of the chromosomes. Although the terms "chromosome" and "gene" are commonly used as if they were synonymous, they aren't. Genes are elements of hereditary information; chromosomes are the boxes they come in. There are estimated to be 35,000 human genes. And if you figure in that genes constitute only an estimated 30 percent of the genome (the rest is intergenic DNA of unknown, and possibly no, function; and 95% of the DNA *within* a gene likewise has no known function), then it follows that a lot of DNA has to be transmitted faithfully every time a cell divides. The DNA accomplishes this by condensing into a manageable number of chromosomes. And obviously there are a lot of genes on each chromosome.

Since the primary function of chromosomes is to get the DNA through cell division smoothly, it doesn't really matter whether there are ten, twenty, or fifty of them. Any number in that range will do well. Consequently, we find that nearly all mammals have between ten and fifty, and in general, closely related species have similar numbers of chromosomes.

Chromosomes can be studied by examining cells (usually white blood cells) and isolating those that are about to divide. They are the ones whose chromosomes are visible. Ordinarily, only a very small percentage of cells is undergoing division at any point in time; however, technological advances in cell biochemistry now permit routine cultivation of blood cells in test tubes, one chemical stimulating them to divide and another preventing them from completing cell division. Thus the proportion of cells whose chromosomes can be studied is much higher than would otherwise be the case.

When humans were thought to have forty-eight chromosomes, it was of interest that chimpanzees had the same number. When humans were found actually to have only forty-six chromosomes, it prompted a recount of chimpanzee chromosomes. But the number remained forty-eight.

Until 1971, each chromosome could be studied only as a uniformly stained silhouette under a microscope; consequently, it was quite difficult to tell one from another. In that year, however, another series of biochemical advances permitted the observation of a characteristic series of bands on each chromosome. Nowadays, a competent cytogeneticist can recognize any specific human chromosome immediately and can often identify any small part of a chromosome.

But if you want to have some fun with a cytogeneticist (as more than one anthropologist has bragged), give them a sample of ape blood, and forget to mention it's not human. Cytogeneticists are trained to identify the human chromosomes and to apply that knowledge in the identification of cancers, which are often associated with broken and reorganized chromosomes. It requires a great deal of thought and skill to recognize that the chromosomes of an ape are those of another species.

Each of the twenty-three pairs of human chromosomes has a nearly identical counterpart in the apes. Perhaps a half-dozen have undergone minor structural changes; but if you can recognize the standard pattern of bands on human chromosomes (known as G-bands), you can recognize chimpanzee chromosomes. It is not known precisely what the bands are, but they can easily be generated in many ways, the most common being to expose the chromosomes to a mild en-

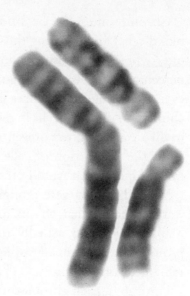

Figure 3. Human chromosomes are almost identical to those of chimpanzees in their fine structure, but there are characteristic differences. The large chromosome on the left is a human chromosome 2, which has no single counterpart in the chimpanzee; rather, its pattern of light and dark bands is matched by two smaller chimpanzee chromosomes.

zyme treatment, which breaks down some of the protein complexes in the chromosome packaging. This permits a stain to bind preferentially to certain areas, creating arrays of alternating stained and unstained regions—bands. And those arrays are nearly identical in humans, chimps, and gorillas.

The principal exception is human chromosome 2, the second largest (they are numbered according to their relative size). You can look at the chromosomes of a chimpanzee forever, and you will never see the large pair known as human chromosome 2. What you will see, however, are two small pairs of chromosomes that when joined end-to-end produce a dead ringer for human chromosome 2 (fig. 3).

Apparently, a fusion occurred in the human lineage, creating chromosome 2 and reducing the count from twenty-four pairs to twenty-three pairs. How do we know it wasn't a fission of chromosome 2 in the chimpanzee lineage? Because we find those two chromosomes in more distant relatives—in fact, as distant as the baboon. That tells us that the ancestral form of the chromosome was as two pairs and the novelty was a fusion in the human line.

Is that what "makes us human"? Yes and no. "Yes" in a narrow, diagnostic sense—given the chromosomes from a cell of any species,

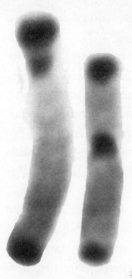

Figure 4. Unlike humans, chimps and gorillas have darkly staining regions at the tips of their chromosomes.

if you find chromosome 2 in it, it's from a human. "No" in a functional sense—the fusion isn't what gives us language, or bipedalism, or a big brain, or art, or sugarless bubble gum. It's just one of those neutral changes, lacking outward expression and neither good nor bad.

The human and chimpanzee chromosomes also differ in a subtle but reliable way under a slightly more complex treatment known as C-banding. C-banding doesn't tell you much, for only a few small regions are stained under this procedure, so most chromosomes look pretty much the same. C-banding seems to mark specifically a few chromosomal zones containing highly redundant "junk" DNA sequences, and in the human the characteristic zones are at the middle or centromere of each chromosome; slightly below the centromere on chromosomes 1, 9, and 16; and along most of the Y chromosome.

And we are the only species with such a pattern. If you look at the chimpanzee's cells using the identical procedure, you find the centromeric bands readily enough, but the regions of chromosomes 1, 9, 16, and Y are simply not there. Not only that, but you will see something entirely unfamiliar—bands at the tips of nearly every chromosome (fig. 4). That is something you'll see in the gorilla as well;

it has been seen in every chimpanzee and gorilla studied but in no human or orangutan.

What is the cause of the bands on the tips? Something trivial, the emergence of yet another functionless class of DNA, which somehow managed to "colonize" the ends of the chromosomes. This class of DNA packs itself more densely into the chromosome than the rest of the DNA does and absorbs more stain. Once the cytogeneticist suspected we had played a joke by handing in a chimpanzee blood sample, a C-band analysis would give it away immediately.

The story is recursive. We find ourselves to be genetically very similar to chimpanzees and yet diagnosably different. Not terribly different from the conclusions we can draw from comparing anything else between chimpanzees and humans—hair, organs, skin, muscles, bones. So the paradox of our great genetic resemblance and notable physical dissimilarity seems less paradoxical.

There is a real paradox, however, that bears considering. We know, on the one hand, that genetic changes are at the root of the visible differences between humans and apes. Franz Kafka's short story "Report to an Academy" gives the reflections of an ape who has become human by virtue of living in Europe and ponders the life he now leads and the one he has left behind. Kafka was using the ape-into-human as a metaphor for the assimilated European Jews in a Christian world, in between the Dreyfus Affair and the Holocaust. But no manner of upbringing will convert an ape into a human; they are immutably, constitutionally different, and programmed that way in their DNA.

And yet when we compare their DNA, we are not comparing their genes for bipedalism, or hairlessness, or braininess, or rapid body growth during adolescence, or the deposition of body fat in the hips and breasts of adult females, or beards in males, or prominent nasal regions. We're comparing other genes, other DNA regions, which have either cryptic biochemical functions or, often, no known function at all.

It's the old "bait and switch." The genes we study are not really the genes we are interested in. We have, sadly enough, exceedingly little knowledge on how a body is put together from genetic instructions. We construct genetic maps principally of its breakdown products—diseases—which are important, but which comprise a very different kind of genetic knowledge. We refer to the gene for cystic fibrosis, the gene for Huntington's chorea, the gene for Duchenne muscular dystrophy, but those identifications are misleading. There are no genes whose function is to give us diseases, there only genes that can be most easily identified by what happens to an unfortunate person who lacks a fully functional copy. This becomes the target of the search (and often the raison d'être for the funding of the research program), but leaves us little knowledge in two crucial areas: first, the genetics of "normalcy"; and, second, the physiology of anatomical development and how it varies among species.

One hypothesis advanced in the 1970s is that there are two kinds of genes, those that code for the biochemical minutiae we are able to study genetically, and those that code for our bodies. The apparent paradox of genetic near-identity, coupled with physical dissimilarity, would seem to be resolved if we postulate that small changes in the first class of genes lead us to the Ivory Soap conclusion, while small changes in the second class of genes lead to major physical changes, the kind that enable you to tell Jane Goodall the person from Flo the chimpanzee at fifty paces.

Once we recognize, however, that both the genetic similarity and the physical difference are somewhat overstated (consider how much easier it is to tell Jane Goodall from Elsie the cow, Charlie the tuna, and Rikki-Tikki-Tavi the mongoose!), there seems to be little reason to imagine two kinds of genes in there. We learn that we are similar to, but invariably distinguishable from, our closest relatives. It couldn't really be any other way, given the fact of evolution.

THE CENTRAL FALLACY OF MOLECULAR ANTHROPOLOGY

The argument I am criticizing here is the one that begins with our unimpeachable genetic similarity to chimpanzees and concludes that we are therefore "nothing but" chimpanzees genetically. In fact, the data have been around for a surprisingly long time, and the fact that it took decades to make this simple, direct deduction might alone suggest that it is specious.

Blood tests used in the 1960s to study the similarity of human and ape proteins were actually little modified from tests developed around 1900 by an English serologist named George Nuttall. By the 1920s, the blood reactions of human and ape sera were well known. For example, journalist and critic H. L. Mencken edited a popular periodical called the *American Mercury,* which featured essays and stories by scientists and other literati. And shortly after the celebrated trial of John T. Scopes (convicted of teaching evolution in Tennessee in 1925), the magazine ran an article called "The Blood of the Primates." And it was quite clear about the results: "[T]he horse and donkey, animals of so intimate a relationship that the one can, as everyone knows, fecundate the other, show a greater difference than do man and the apes. In other words, it is proved that the sanguinity of the horse and donkey, which are capable of hybridization, is less close than the kinship of *Homo sapiens* and the anthropoids." The article failed, in spite of the data, to draw the conclusion that we "are" apes. That inference—which I shall call The Central Fallacy of Molecular Anthropology—would have to wait for the molecularization of biology in the 1960s.

Ultimately, the fallacy is not a genetic but a cultural one—our reduction of the important things in life to genetics. Geneticists are among the most frequent perpetrators of this fallacy, and of course it is in their interests to perpetuate it. In the 1960s, when it became possible to make direct comparisons between the proteins of different species, biochemist Emile Zuckerkandl was impressed by the fact that only one site in the 287 amino acids of hemoglobin could be found to differ between a human and a gorilla. So impressed, in fact, that he proclaimed that "from the point of view of hemoglobin structure, it appears that gorilla is just an abnormal human, or man an abnormal gorilla, and the two species form actually one continuous population."

Well, of course, as the distinguished paleobiologist George Gaylord Simpson was obliged to point out in quick response, a gorilla is not an abnormal man; it's a gorilla. A man is not an abnormal gorilla;

he's a man. And gorillas and humans do not form a single population, any more than gorillas and daffodils do. And furthermore, if hemoglobin seems to be telling you anything else, it's lying.

When I met Simpson, the patriarch of mammalian evolution studies, toward the end of his life in the early 1980s, he was still mystified by the combination he saw of ignorance and arrogance on the part of students of molecular evolution. He had led a fuller and more prolific life than most—he told me he had been in the army hospital the day General Patton famously slapped the soldier; he had survived a tree crushing his legs while on a fossil expedition in Brazil in 1956— and now continued to write books about evolution while in retirement in Tucson. His stature in the field was such that, as a graduate student, I developed a stutter whenever I spoke to him on the phone, but in person, he was a reserved and humble old man. He still wore the neatly trimmed beard he had always worn, now white with age. Sipping the two martinis that constituted his lunch, he pronounced with a shrug on deducing the sameness of humans and gorillas from the sameness of their hemoglobin: "It's just plain dumb!"

And that from the man who had been preeminently responsible for integrating modern genetics into evolutionary theory at midcentury.

Hemoglobin, of course, doesn't tell lies or truths. It indicates simply a set of data and interpretations. And in this case the interpretation was a facile one. Why on earth should hemoglobin be a microcosm for the relations of humans and gorillas? Surely we are not strictly reducible to our blood, any more than we are reducible to our arm, our toenails, or our phlegm. If we are similar but distinguishable from a gorilla ecologically, demographically, anatomically, mentally—indeed, every way except genetically—does it follow that all the other standards of comparison are irrelevant, and the genetic comparison is transcendent? More likely, it should imply that the genetic comparison is exceptional, and that it is interesting for that very reason.

Suggesting that the relationships of our *blood* are the relations of *us,* that *we are our blood,* is simply a metaphoric statement, technically called metonymy, the substitution of a part for the thing itself. Perhaps it is resonant with us because our brains evolved to make meta-phoric connections, and because blood is such a symbolically pow-erful substance. Blood is, after all, a metaphor for heredity itself.

But this metaphor is not literally true; and science is supposed to be about literal truths, not literary ones.

There is one sense in which we can acknowledge that we are apes—phylogenetically. We fall phylogenetically within the group consti-tuted by the "great apes"—those large-bodied, tailless, flexible-shouldered, slow-maturing primates—which also includes chimpan-zees, gorillas, and orangutans. And indeed, we are more closely related to the chimpanzee and gorilla than they are to the orangutan. This implies that that category of "great ape" is artificial, for it comprises species that are not one another's closest relatives; or rather, it excludes a member of the group of relatives—namely, us.

We are excluded by virtue of the fact that we have diverged rela-tively rapidly from the classic ape form and mode of life and have evolved to fill a different niche, that of an "ape" that is bipedal and culture-reliant. So although we fall within the great apes genetically and are recently descended from them biohistorically, we are never-theless different from them by virtue of having evolved a very large number of readily observable specializations or novelties. We have left them in our wake, so to speak, and the apes seem to resemble one another more than they resemble us, because they didn't develop the things we did.

But does this mean that we *are* apes, as Oxford professor and ge-netic enthusiast Richard Dawkins has argued?

Consider a different group of animals: say, sparrows, crocodiles, or turtles. Two are crawling, green-scaled reptiles, and one is a bird. As it happens (like humans and gorillas relative to orangutans), the sparrow and crocodile are more closely related to each other than ei-ther is to the turtle. The two reptiles are more similar physically to one another, but they are similar simply because the birds, which originated from a group of reptiles, developed a set of specializations and flew away, leaving the green scaly creatures behind. Since the kind of reptile from which the birds originated was related to the crocodile, it happens that the sparrow and crocodile, a few hundred million years later, are more closely related than either is to the turtle.

But does that mean that the sparrow is a reptile? No, it means it is closely related to reptiles; its group is subsumed by the reptiles—but if the word "bird" is to have any meaning, it must be a different meaning from the word "reptile." Birds are birds; reptiles are a different kind of group, unified by not having evolved the specializations of birds.

Closer to home, my family is distinct from me, although it subsumes me. In one sense, I am a part of my own family. But in a more immediate sense, my love of my family does not necessarily imply a narcissistic love of myself, for it refers to "my family, excluding me." Sometimes the exclusion is what makes a statement sensible.

We can go even further back in biohistory. Several hundred million years ago, a group of fish developed specializations of their limbs that enabled some of their descendants to venture out of the sea and on to land. A living representative of that group of fish is the famous "living fossil," the coelacanth. Since all living land vertebrates (or tetrapods) are descended from that particular group of fish, it follows that if you compare a coelacanth, a human, and a tuna, the closest relatives are the coelacanth and the human, not the two fish.

Let's return to the apes. The argument is that humans are apes, because we belong to the group that produced chimpanzees and orangutans; and because we are more closely related to some apes than those apes are to other apes, we fall within that category. That argument is structurally identical to the argument that humans are fish, because we belong to the group that produced coelacanths and tuna, and because we are more closely related to some fish (coelacanths) than those fish are to other fish (tuna). We fall (by virtue of being tetrapods) within that category as well.

In other words, we are apes, but only in precisely the same way that we are fish (fig. 5).

Doesn't seem quite so profound now, does it?

That's because this is not so much a revelation about our basic natures as about how we name zoological groups. As constituted by a group of related species that a subset has evolved away from, the

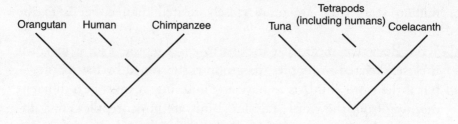

Figure 5. Humans fall within the phylogenetic category of apes, although we have diverged physically from them (physical divergence is represented here by broken lines); likewise, tetrapods fall within the phylogenetic category of fish, although they have diverged physically from them. Thus humans are phylogenetically apes only in the same sense that we are phylogenetically fish.

species that remain unchanged will constitute a "paraphyletic" category. They may look similar, but they are defined on the basis of lacking the specializations of the other group—tetrapods in the case of paraphyletic fish, birds in the case of paraphyletic reptiles, and humans in the case of paraphyletic great apes.

What appeared at first to be a revelation about our animal nature is instead a revelation about how we box up nature to make sense of it. Making sense of the world through classification is a fundamentally human act, and each human group does it in its own way, and thus imposes structure upon the world for itself.

Modern science classifies animals by two criteria, descent and divergence. This creates confusions such as the occasional paraphyletic category, in which divergence takes precedence over proximity of descent. Some biologists prefer to classify by descent only and would thus recognize only closest relatives, burying the categories "great apes," "reptiles," and "fish." Ultimately, this is a philosophical decision, is unresolvable by recourse to data, and displays the variation of ideas about classifying among scientists. They are a human society, and like other human societies, they make sense of their world by organizing it into groups.

The ancient Hebrews classified for a different reason and used different criteria. Their goal was to distinguish symbolically clean from unclean, and their criteria were where the animal lived, and how it moved. For this reason, shellfish and pigs were prohibited but chickens and locusts were not. Leviticus 11 and Deuteronomy 14 give us the following categorization:

> Living on the Ground
> > Animals with hooves
> > > Split hoof and chews cud [EDIBLE]
> > > > *ox, sheep, goat, hart, gazelle, roebuck, wild goat, pygarg (ibex), antelope, mountain sheep*
> > > Whole hoof and chews cud [INEDIBLE]
> > > > *camel, hare, rock-badger*
> > > Split hoof but does not chew cud [INEDIBLE]
> > > > *pig*
> > Animals with paws [INEDIBLE]
> > Animals that creep or swarm [INEDIBLE]
> > > *weasel, mouse, lizard, gecko, crocodile, chameleon*
> > Animals that travel on belly [INEDIBLE]
> > Animals with many feet [INEDIBLE]
> Living in the Water
> > Animals with fins and scales [EDIBLE]
> > Animals without fins and scales [INEDIBLE]
> Living in the Air
> > Animals that fly
> > > Clean birds [EDIBLE]
> > > Unclean birds [INEDIBLE]
> > > > *eagle, vultures, osprey, glede, falcon, kite, raven, ostrich, hawks, sea-mew, owls, cormorant, pelican, heron, hoopoe, bat*
> > Animals that swarm, with wings and jointed legs [EDIBLE]
> > > *locust, grasshopper, cricket*
> > Animals that swarm, with four feet [INEDIBLE]

One of the most widely studied classification systems cross-culturally is that of kin. We give the same name, "aunt," to four very different relatives, for example: our mother's sister, our father's sister, our father's brother's wife, and our mother's brother's wife. The latter two are not even genetic relatives, and of those, at least the first shares our family name. There is no necessary reason to call them all "aunt." The fact that we call such different people the same thing is not rooted in nature or genetics, but rather in an attempt by our ancestors to make sense of the complex social relations they had to deal with, by giving names to classes of people. As a result of the conventions of categorization and language, we draw a distinction, for example, between "uncles" and "aunts," but in English we do not draw the same distinction (by sex) for "cousins." In fact, we don't even have a word by which to designate our cousin's spouse.

CLASSIFICATION AS A CULTURAL ACT

There is a strong analogy to be made between classifying relatives and classifying species. In both cases we are taking some aspect of nature, either the state of kinship or the state of evolutionary populations, and imposing a cultural system of nomenclature upon it. In the case of our relatives, we rarely question why our uncle's spouse should be a named relative but our cousin's spouse should not be, or why we should call people "aunt" who stand in such different relationships to us.

But surely the situation is different for classifying species, where the order is directly derived from nature as a result of evolution.

Actually, though, it isn't. In the social world, there are blood relatives of different degrees, relationships derived from nature, which our system of kinship amplifies but does not perfectly reflect. Likewise with species, where the natural relationships often encode unnatural information, most notably, as we have seen, paraphyly.

There are many arbitrary decisions that are imposed upon nature in the act of classifying. At root is a classic problem that exists above the species level: Are two animals that look somewhat different from one another to be placed in different genera or in different families? There is, in fact, no answer given to us by nature. The best answer nature gives us is just at the species level, where parts of the same species share in a common reproductive community and parts of different species do not.

In the "system of nature" invented by Linnaeus, every species of animal was given a double name (such as *Homo sapiens*) and a unique place in a four-tiered, nested hierarchy—genus, order, class, and kingdom. Many intervening categories have since been added since Linnaeus's time, and many changes have been made to his system. But in the most basic ways, his work stands. For example, in the 1758 tenth edition of his *Systema naturae,* from which modern zoological classification dates itself, Linnaeus decided that the natural place of humans was within a class he called Mammalia.

Now, every biology student knows that humans are mammals. It would seem like a simple fact of nature, quite the opposite of the cultural situation described earlier. But recent scholarship has shown that calling humans "mammals" is in fact a strongly cultural statement. Why did Linnaeus focus on our mammae? (Not, as a student once told me, "because he was a guy"!) Why is milk so damn important in the great scheme of things that we should take our very name on that basis? Couldn't we come up with the same group of animals using a different criterion, and so why don't we?

For example, Aristotle over two thousand years ago called land animals Quadrupedia (four-legged) and divided them into those that lay eggs and those that give birth to live offspring. And "four-legged creatures that give birth to live offspring" gives you basically the same constellation of animals, with a few exceptions, like the duck-billed platypus.

Mammals actually have a *lot* of features that distinguish them from reptiles, amphibians, fish, and birds. Hair, for one thing. And some scientists in the eighteenth century did call this group "Pilosa," or hairy things. But Linnaeus called us mammals, based on an anatomical feature that's only functional in half of our species, and then only occasionally.

So why did he do that?

He had already made a reputation by classifying plants according to their sexual anatomy. This might have been simply a way of bringing animals into that same paradigm. But there is no compelling reason why animals and plants should necessarily be subject to the same criteria.

It turns out actually to have been a political gesture. As historian Londa Schiebinger points out in her book *Nature's Body,* in the 1750s there was a major controversy going on about wet-nursing. The European middle and upper classes at that time mostly sent their babies off to be breast-fed by poor women in the countryside, rather than nursing their own babies themselves. Linnaeus was active in the movement that opposed this practice. In fact, he wrote a book on the virtues of breast-feeding your own children, asserting that it was natural for mothers to do so, and therefore wet-nursing was unnatural and bad. Up to that time, he had been calling mammals simply Quadrupedia, like Aristotle. Now he called mammals Mammalia, using his scientific classification to make a point. He was saying that women are designed to nurse their own children, that it is right, and that it is what your family should do.

The point of all this is to show that what a biology student takes for granted as a fact of nature—that we are essentially definable as a lactating species—was actually historically produced: it is an eighteenth-century political stand encrypted into biology. The fact is, of course, that mammals are a natural unit, which can be defined by nursing, but being natural doesn't make breast-feeding an objective category, simply "out there" to be discovered and adopted as a scientific identifier.

It is not obviously the case that breast-feeding is the key feature that makes us mammals, any more than having a single bone in the lower jaw (which all Mammalia have, and *only* Mammalia have) is the key feature that would make us "One-bone-in-jaw-malia." There's more here than nature.

To classify is to be human: mythical Adam's first task was to classify. We rarely question the categories within which we work because they seem so natural to us; we are accustomed to them. But they are constructions of human minds, the products of social history. That they seem natural or "given" or "out there" is a testimony to their power over us and the difficulty we have dropping them. While it may seem arcane to worry about the existence of paraphyletic categories, this cultural drama is played out on a larger stage when we attempt to come to grips with racial categories.

Three

HOW PEOPLE DIFFER FROM ONE ANOTHER

THE MOST PROMINENT, OR at least newsworthy, of our seemingly natural categories is race. We tend to divide people into a few large categories: white, yellow, black, red, or the more politically correct Europeans, Asians, Africans, and Native Americans. And we tend to perceive facial features that stereotypically follow this division: light skin, almond eyes, kinky hair, beaked nose. In fact, these categories do not reflect any natural division of the human species and have principally been the construction of social history.

Teaching that racial categories lack biological validity can be as much of a challenge as teaching that the earth goes around the sun must have been in the seventeenth century—when anyone can plainly see the sun rise, traverse a path across the sky, and set beyond the opposing horizon. How can something that seems so obvious be denied? Of course, that is the way all great scientific breakthroughs appear, by denying folk wisdom and replacing it with a more sophisticated and analytic interpretation of the same data.

It is hard to imagine that the order we perceive in nature may not be nature's order at all but our own order superimposed upon it. Nevertheless, that is one of the great lessons of modern anthropology, demonstrated specifically in the history of the classification of peoples and consistently reinforced by genetic data.

RACE AS A SCIENTIFIC ORGANIZING PRINCIPLE

The ancients knew a rather limited world, but they were certainly impressed by the ways in which people from different places looked different from one another. Writing in the fifth century B.C., for example, Herodotus noted that the people of Colchis and Egypt had dark skin and wooly hair, and that the people of Ethiopia were said to be the tallest and handsomest in the world. To Herodotus each land had its own characteristic peoples. But there was no hint of the existence of any small fundamental number of natural human kinds.

The ensuing centuries saw little change in this unsystematic conception of human differences. Science itself perceived little unity in the natural world until the seventeenth century, beginning with Galileo and ending with Newton, identified the unifying physical principles governing matter and motion.

This engendered a profound change in the approaches of scholars to the natural world. Where once they had seen only individual phenomena and interpreted them as evidence of God's bounty, they now began to see general principles at work and interpreted them as evidence of God's parsimony. Chaos would be replaced by order, disarray by system. To understand the system was to know God's plan.

Thus it was that a French physician and traveler named François Bernier contributed something novel and original to the study of human differences—the idea that humans could be grouped systematically into a relatively small number of classes. Proclaimed Bernier, "there are four or five species or races of men in particular whose difference is so remarkable that it may be properly made use of as the foundation for a new division of the earth."

Bernier's first group encompassed Europe, North Africa, the Near East, and India. His second was sub-Saharan Africa, characterized by people with thick lips and flat noses, dark skin, scanty beards, and noteworthy aspects of their hair ("which is not properly hair, but rather a species of wool, which comes near the hairs of some of our dogs") and mouth ("teeth whiter than the finest ivory . . . their tongue and all the interior of their mouth and their lips as red as coral"). The third encompassed the Asians, who were "truly white; but they have broad shoulders, a small [flat] nose, little pig's-eyes long and deep-set," and a scanty beard as well. His fourth comprised the Lapps of Norway, "little stunted creatures with thick legs, large shoulders, short neck, and a face elongated immensely . . . , they are wretched animals," and somewhat bearlike as well. As to the native Americans, Bernier did "not find the difference sufficiently great to make of them a peculiar species different from ours." And finally, "the blacks of the Cape of Good Hope seem to be of a different species to those from the rest of Africa. They are small, thin, dry, ugly, quick in running, passionately fond of carrion which they eat quite raw, and whose entrails they twine round their arms and neck . . . ; drinking sea-water . . . , and speaking a language altogether strange, and almost inimitable by Europeans."

Where formerly there had been diversity of form, Bernier now found unity. No longer was local variation significant, rather, there were just a few basic forms—species, or races.

Of course, the terms then did not have the same formal definitions as they do now. Thus, when Bernier calls the Khoisan "a different species," he is not working with the biological definition we now employ. He means merely a population that looks different (*really* different).

Perhaps these differences reflected different origins—separate creations, as a later school, known as "polygenists," would come to argue. Not surprisingly, this academic position reached its zenith during the American Civil War. Or perhaps the races represented the different offspring of the biblical patriarch Noah. The Bible sort of suggests this in Genesis 10, giving a list of descendants of Noah's sons, which are also the names of cities and peoples.

And by the first century A.D., that had indeed become a widespread interpretation. Noah's son Ham was the father of the Africans, Shem was the father of the Asians, and Japheth the father of the Europeans. The gloss was put on the story in an influential popular work, *Antiquities of the Jews,* by Flavius Josephus—a renegade from the Judean army crushed by Rome in the year A.D. 71. Josephus was concerned to explain away the Jewish War, in which God's chosen people were soundly thrashed, and to recount the ancient history and customs of his people to their conquerors. Thus, he summarizes the Jewish Bible, including not simply the literal stories but the general extracanonical understandings of them as well.

In this account, after Noah's death, God tells his three sons "to send colonies abroad, for the thorough peopling of the earth." But they disobey and hang together, under "the suspicion that they were therefore ordered to send out separate colonies, that, being divided asunder, they might the more easily be oppressed." Two generations later, Ham's grandson Nimrod builds the Tower of Babel, which God not only destroys, but also curses the builders by giving them different languages. Now, says Josephus, "they were dispersed abroad, on account of their languages, and went out by colonies everywhere; and each colony took possession of that land which they light upon, and unto which God led them; so that the whole continent was filled with them, both the inland and the maritime countries."

The geographical details in Josephus are sometimes a bit obscure, but it is not difficult to see how easy it was to equate Europeans with Japheth, Asians with Shem, and Africans with Ham. Of course, Josephus and other ancient writers were not familiar with southern Africa or east Asia, not to mention America or Australia.

But by the seventeenth century, European scholars were not only aware of the diverse peoples of the most remote regions of the world, they appreciated that humans everywhere could hybridize freely. (Sailors had proved that, not scientists.) Given, then, that we all constituted a single biological entity, what were that entity's basic components?

Europeans freely speculated on the divisions; thus a tract from 1721 finds

> five sorts of men: the white men, which are Europeans that have beards; and a sort of white men in America (as I am told) that only differ from us in having no beards; the third sort are the Malatoes, which have their skins almost of a copper colour,

small eyes, and straight black hair; the fourth kind are the Blacks, which have straight black hair; and the fifth are the Blacks of Guiney, whose hair is curled, like the wool of a sheep. . . .

It is always hard to know why certain things catch on and others don't—why we have VHS and not Beta; compact disks and not quadraphonic stereo; Honda Civics and not Ford Edsels; Christianity and not Mithraism. In this case, it is not hard to see a self-serving harmony between a neat and simple classification of people and the economic and political systems emerging in Europe. Focusing ever more keenly on the exploitation (from economic dependency to displacement, enslavement, and even genocide) of large numbers of diverse peoples, the "spirit of the times" could make this system seem very sensible. If many different people were all basically the same, it would be natural to treat them in a common fashion; and so the relations of "European" and "African" could come to subsume those of the specific colonial nations or administrations and the specific indigenous peoples they abused. The ease with which this classification fitted into the prevailing political and economic climate made it seem almost ordained by natural law.

And what is zoologically natural was codified by Linnaeus.

Writing only a few decades after Bernier, Linnaeus began the classification of animals in his *Systema naturae* with four kinds of *Homo:* white Europeans, red Americans, yellow Asians, and black Africans. He had not yet developed the formal binomial nomenclature that would name species in two parts (*Homo sapiens* for humans, for example). Furthermore, his system does not precisely coincide with Bernier's. For example, the Lapps held a romantic fascination for Linnaeus as innocent Scandinavian "noble savages"—but certainly not as a formally recognizable entity separated from other Europeans. Nevertheless, he incorporated Bernier's general innovation into the scientific classification of humans and even color-coded them for your convenience.

The tenth edition of Linnaeus's *Systema naturae* (1758) included such significant changes as substituting "Mammalia" for "Quadrupedia" and "Primates" for "Anthropomorpha," as well as introducing *sapiens* as the second half of our own species' name. But he did not deviate from the decision over twenty years earlier on the four geographical human tints.

Once again, though, we are reminded that these ideas are not quite the same as our own when we see that Linnaeus also introduced a fifth subspecies, not of geographic origin, called *Homo sapiens monstrosus*. This was a geographical grab bag, including giant Patagonians, cone-headed Chinese, flat-headed Canadians, and one-testicled Hottentots.

Linnaeus's first four color-coded subspecies, however, are the ones that became the official, scientifically accepted subgroupings of man (at least until Captain Cook's voyages to Oceania and Polynesia later in that century). And it is worth examining the structure of these subgroupings, for they reveal just how unlike the other species of animals Linnaeus treated this one. He is justly remembered for the radical step of subsuming humans into the natural classification of organisms, but did he treat humans the same way he treated anteaters?

Actually, Linnaeus rarely uses the concept of the subspecies. He divides only two other groups of mammals (dogs and sheep) that way. And his presentation of the human subspecies has a distinct structure to it: three adjectives corresponding to color and temperament; a brief description of the appearance; three adjectives describing the personality; a mention of their bodily covering, and a single word encapsulating their legal system (see table).

Linnaeus's training as a physician led him to interpret these differences among the four subspecies in a medical fashion. In particular, his descriptions of the four temperaments (*cholericus, sanguineus, melancholicus,* and *phlegmaticus;* or irascible, vigorous, melancholy, and sluggish) correspond to the four Hippocratic humors of which the body was thought to be composed. These were yellow bile or *chole,* blood (*sanguis*), black bile or *melancholia,* and phlegm.

In this view of health, the body was seen as normally being in a state of balance among these four substances. Variation of behavior, or attitude, or personality was interpreted as an overbalance of one

Linnaeus's Classification of the Species *Homo Sapiens* in *Systema Naturae* (1758)

	American	European	Asian	African
Color	red	white	yellow	black
Temperament	irascible, impassive	vigorous, muscular	melancholy, stern	sluggish, lazy
Face	thick, straight, black hair; broad nose; harsh appearance; chin beardless	long blond hair, blue eyes	black hair, dark eyes	black kinky hair, silky skin, short nose, thick lips, females with genital flap, elongated breasts
Personality	stubborn, happy, free	sensitive, very smart, creative	strict, contemptuous, greedy	sly, slow, careless
Covered by	fine red lines	tight clothing	loose garments	grease
Ruled by	custom	law	opinion	caprice

or more of the four humors. Each humor was in turn associated with specific body organs: the gall bladder, heart, spleen, and pituitary. And of course, with the continents—America, Europe, Asia, and Africa. What Linnaeus thus seems to have done here is to have encoded the peoples of the earth as the parts of a macrocosm of a contemporary human body, each subspecies from each continent governed by one of the four humors and organs.

This simply shows that Linnaeus was interpreting human diversity as well as he could scientifically with the conceptual apparatus of his training in the culture of eighteenth-century Europe. His subspecies are idealizations—defining Europeans zoologically as having long blond hair and blue eyes, for instance, is an understandable absurdity coming from a provincial Swedish biologist, but it is hardly realistic. It reflects a decidedly premodern, essentialized view of the world, in which qualitative distinctions are assigned to each category on the basis of symbolic distinctions, not real ones.

Essentialism is a mode of thought that we associate in biology with the Greek philosopher Plato. Plato's concern in many of his writings was what the ideal world would be like, and his writings on nature tend to eschew empirical reality for a deeper reflection on deeper realities. In his famous Cave Allegory, Plato wants us to look beyond what we actually see (the shadows on the cave wall) to deduce what we cannot see, but that which is the ultimate cause of the forms we perceive. It takes reason, says Plato. While seductive in this form, the allegory turns out to be a problem for science. It asks us to walk away from our sensory experience, from data, from reality, and to think instead about an imaginary hyperreality. The problem is that the core of science lies in the connections between data and explanations; explanations are cheap, but good explanations that permit themselves to be judged on the basis of their fit to the real world—the shadows we see—are precious.

In Linnaeus, we see the problem at its most glaring. When Linnaeus, for example, *defines* Europeans as having long blond hair and blue eyes, he is obviously not *describing* them—he is idealizing them. He is describing an imaginary man whose "shadows" he believes to constitute the diverse faces and bodies of real Europeans. We shall see the fallacy of essentialism crop up repeatedly in different biological studies throughout the centuries, but always bearing the hallmark of denying the importance of real variation and focusing instead on an imaginary uniformity.

This would be merely a bit of historical esoterica were it not for one important point. Linnaeus had indeed discovered a fundamental truth about life on earth: it comes packaged in nested groups of increasingly general similarity to one another. Thus, species cluster into genera, like Linnaeus's *Homo sapiens* and *Homo troglodytes* (a mistake on his part, as noted in chapter 1). Genera cluster into orders: his *Homo* and *Simia* are both Primates. Orders cluster into classes, as his Primates and Ferae (or what we would now call Carnivora) do into Mammalia, and Mammalia and Aves do into the kingdom Animalia. What this represents is of course the patterns of evolutionary descent, with lower-level categories sharing more recent ancestry with one another—at least to a first approximation.

Linnaeus did not know the explanation for the pattern. All he knew was that the pattern was there, and it became obvious to the rest of the scientific community overnight, so that Linnaeus's classification became *the* way to do biology. He supervised 186 doctoral theses, an output so prodigious as to be difficult to believe nowadays (to put this into context, Sherwood Washburn, arguably the leading American biological anthropologist of the latter half of the twentieth century, told me he supervised about 40 doctoral theses in the course of a full and distinguished career at Columbia, Chicago, and Berkeley).

Linnaean biology, briefly put, *became* biology. To study nature was to study it his way, to classify it; to identify or infer the genera within each family (the family as a classificatory level was introduced by his successors), the species within each genus, and (at least in the case of humans), the subspecies within each species.

A contrary view, however, was offered by the comte de Buffon, an erudite French *philosophe,* in his widely read *Natural History, General and Specific,* also published during the middle of the eighteenth century. Buffon famously rejected the entire Linnaean classificatory system, for an historically curious reason. According to Buffon, these nested categories directly implied a common ancestry for the species contained within them, and he knew that could not be true! As he wrote with extraordinary insight, "The naturalists who establish the families of plants and animals so casually do not seem to have grasped sufficiently the full scope of these consequences, which would reduce the immediate products of creation to as small a number of individuals as one might wish."

He was absolutely right; they did not fully grasp the consequences of what they were doing, not until Darwin explained it a century later. Buffon in any case defiantly rejected Linnaeus's classificatory system, and his work was influential with the public (although less so with the academics; Linnaeus was, after all, right).

In his survey of the varieties of the human species, Buffon consequently takes a very different approach from that of Linnaeus, not carving the species into subspecies, but simply surveying the diversity of form and behavior "out there." His descriptions are often quaint and ethnocentric, but they are quite different in structure from those of Linnaeus. Most importantly, however, Buffon recognized a pattern of variation in the human species quite different from that promoted by Linnaeus: namely, continuity, the absence of discrete boundaries between any distinct formal categories of humans. "On close examination of the peoples who compose the black races," says Buffon, "we will find as many varieties as in the white races, and we will find all the shades from brown to black, as we have found in the white races all the shades from brown to white."

The two naturalists hated each other. Although both were ennobled late in life, Linnaeus had come from far humbler origins and had known poverty. He was an academic and wrote in scholarly Latin; Buffon was a wealthy landowner who wrote for the educated masses in eminently readable French. Linnaeus found Buffon's writing fluffy and disorganized; Buffon found Linnaeus philosophically unsophisticated and sterile—what good was classifying without comprehending? "The misunderstanding between these two able Naturalists is most injurious to science," their English contemporary Thomas Pennant complained. "The *French* Philosopher scarce mentions the *Suede,* but to treat him with contempt; *Linnaeus* in return, never deigns even to quote M. *de Buffon,* notwithstanding he must know what ample lights he might have drawn from him." Linnaeus eventually acknowledged Buffon by naming a foul-smelling plant after him.

A curiously relevant paradox arises in the general conflict between these two savants of the Enlightenment era. Linnaeus formalizes the divisions among humans, but does not call them "races"—they are, for him, varieties or subspecies. Buffon, on the other hand, denies the value of classification, denies there exist formal subspecies of humans, yet talks at length about "races"—in a very loose and colloquial manner, akin to stocks. The next generation of scholars would take

Linnaeus's approach and apply Buffon's term—converting the formal subspecies of humans into "races."

Whether you called them "races" or "subspecies," the goal was the same: to study humans scientifically, you had to classify them. To classify them was to transcend the shadows on the cave wall, and infer the basic plan of nature, at least insofar as the diversity of the human species was concerned. And thus it was that at the end of the eighteenth century, Johann Friedrich Blumenbach could write that "one variety of mankind does so sensibly pass into the other, that you cannot mark out the limits between them," and then nevertheless proceed to attempt just that. Donning the mantle of Linnaeus, Blumenbach jettisoned the nonbiological criteria Linnaeus had employed and identified five races: Mongolian, Ethiopian, Malay, American, and, his most famous coinage, Caucasian.

PHYSICAL ANTHROPOLOGY AS LINNAEAN SCIENCE

Physical anthropology became the scientific study of racial differences—assuming, of course, that there were indeed races, and that they were formally different. To deny this was, however, to undermine the raison d'être of the field, especially as it developed in American academia in the early twentieth century under the influence of the organizational skills of a Bohemian physician-cum-anthropologist named Aleš Hrdlička, and the pedagogical skills of a classicist-cum-anthropologist from Wisconsin named Earnest Hooton. In laying claim to the scientific study of race as their "turf," Hrdlička (at the Smithsonian) and Hooton (at Harvard) became the authorities on what was and wasn't *really* a race.

The two men could hardly have been more different. Hrdlička, a haughty Central European immigrant, had had a classical medical training and studied with the French physical anthropologist Léonce Manouvrier before settling at the Smithsonian and becoming the expert on the origins of Native Americans. The good-natured Hooton, who came from old Anglo-Saxon stock, breezed into physical anthropology while on a Rhodes Scholarship at Oxford. Appointed to the Harvard faculty in 1913, he established himself as a glib lecturer and witty writer, most especially on the nature of human differences. With titles like *Apes, Men, and Morons; Man's Poor Relations;* and *Young Man, You Are Normal,* his books were widely read, and his opinions were regarded as authoritative. For nearly half a century, virtually anyone who wished to study physical anthropology in the United States came under his tutelage.

Hooton's domain was race.

And what was race? Race was formally "a vague physical background, usually more or less obscured or overlaid by single subjects and realized best in a composite picture." Hardly the stuff of a precise mature science. And yet Hooton would also write sarcastically of those anthropologically illiterate "ethnomaniacs" who spoke casually of a Jewish "race," a Latin "race," or an Irish "race." The proper usage, argued Hooton, should distinguish only each "great division of mankind" on the basis of their physical uniformities. But there was still room for pure and isolated "primary" races and admixed "secondary" races.

The ambiguity existed largely because there were still two different senses of the word "race" concurrently in use. The first, the colloquial use of the word that Buffon had adopted, meant pretty much any group of people with their own name. That, of course, couldn't be scientific. The second was Linnaeus's formalized subspecies, the "scientific" sense of the word "race," even if Linnaeus himself didn't use it. Using both together produced the absurdity of those races within races, some of which had more pronounced biological distinctions and others of which did not. Neither was it clear whether individual people constituted a race, or whether race was something that existed within an individual. Racial identification, argued Hooton, was like a physician diagnosing a disease; one needed to examine the symptoms before making a judgment, because a subject could look white and really be black, or vice versa—thus, *races were in people; people weren't in races.*

The situation was so confused that, as the British polymath Lancelot Hogben would write: "[G]eneticists believe that anthropologists have decided what a race is. Ethnologists assume that their classifications embody principles which genetic science has proven to be correct. Politicians believe that their prejudices have the sanction of genetic laws and the findings of physical anthropology to sustain them."

Enter genetics.

The earliest genetic analysis of the human species was based on the ABO blood group, discovered just after 1900 and surveyed during World War I. Somehow it managed to divide the peoples of the world into "European," "Intermediate," and "Asio-African"; or, essentially, "white" and "other." The curious fact is that if you look at any contemporary textbook today, you will discover that the ABO blood group is now invoked as a demonstration of precisely the opposite point—that there are no discrete divisions of our species possible on this basis.

How could the same genetic data have yielded a division of the human species into distinct genetic races at one point in time and demonstrate that there are no such distinct races at another point in time? Obviously, the answer lies with what the geneticists expected to find at different times. When they expected to find discrete divisions, their data yielded them; and when they expected *not* to find discrete divisions, their data did *not* yield them.

Race, it seems, is *underdetermined* by genetic data.

A more sophisticated analysis of the ABO blood group in 1926 divided the world into seven blood-group racial types on the basis of their proportions of each of the ABO variants: European, Intermediate, Hunan, Indo-Manchurian, Africo-Malaysian, Pacific-American, and Australian. The problem was that the classification was exceedingly unnatural. Since the Poles and the Chinese had the same frequencies of the blood-group genes, they were classified together as belonging to the "Hunan" type; but if there were anything at all to the concept of race, the people of Poland and the people of China more or less had to be in different ones. Likewise, the diverse peoples of Senegal, Vietnam, and New Guinea ended up together as "Africo-Malaysian."

Earnest Hooton at Harvard was dismayed. On the one hand, this was authoritative genetic work. On the other hand, it produced meaningless clusters of people, basically scientific gibberish. Once again, the geneticists had impressive batteries of data but interpreted the data within a fairly naïve and narrow framework, bolstered by the cachet of science.

For example, the blood-group work didn't end with just the classification of races. It also conjured up a biological history for those races. Finding that all populations of the world have mostly type O blood, and that Native Americans are almost exclusively type O, the geneticists reasoned that the Native Americans must represent the most primitive branch of humanity, diverging from the human stock before any other population—in spite of the fact that they appeared to resemble North Asians very strongly, suggesting a very recent common ancestry with them. And as for the rest of the world, the originally pure O race that settled in Europe must have been invaded by the A race from the north, and the B race from the south, resulting in the diverse populations that now exist.

The idea that the human species might be primordially polymorphic, and might have possessed all three variations from its very beginning, was hardly entertained—in spite of the fact that A, B, and O were known to exist in the genes of the apes as well. In an era that valorized purity of race, the idea was incorporated whole-hog into geneticists' scenarios with little critical thought.

Obviously, the geneticists were simply reading their expectations of discrete divisions of the human species into their data. Quite possibly the most bizarre of these studies was performed in the Soviet Union by a hematologist named E. O. Manoilov. If he added a few simple chemicals to a small sample of blood (a powerful metaphor, of course, for heredity) under controlled conditions, Manoilov reported, the blood changed color. Initially, he found that he could discriminate male from female blood—and it even worked for plants, in spite of the difficulties imposed in extracting blood from them. Slightly later, he could distinguish the bloods of different races, notably Poles, Koreans, Russians, and Jews. And as if that were not enough, the blood test could even distinguish the blood of heterosexuals from that of homosexuals.

Certainly, no scientist would have welcomed a blood test for distinguishing race more than Hooton, but he appreciated that sociopolitical categories such as "Jew," "Russian," and "Pole" were the constructions of human history and simply couldn't be distinguishable by the properties of blood. He dismissed the Manoilov blood test in a single sentence in his 1931 text *Up from the Ape,* and the test disappeared from the anthropological literature.

It had a different fate in the genetics literature, however. Thomas Hunt Morgan had a Russian visitor to his laboratory at Columbia translate Manoilov's work and look into it. Charles Davenport, the leading human geneticist in America, put some of his junior colleagues on the sex determination test, and they published "preliminary results" in major journals. Davenport also enlisted a biochemist named K. George Falk to study the Manoilov blood test, and after some encouraging preliminary results, Falk found to his surprise that the blood test also worked on urine. He couldn't explain it, and they abandoned the Manoilov blood test shortly thereafter.

To this day we don't know what the Manoilov blood test, which mysteriously required no blood, was actually testing. But even as it was disappearing from the anthropology literature, it was being uncritically cited in mainstream genetics textbooks. Said one major 1931 college textbook, "According to Manoiloff, the oxidizing process in a certain blood reaction occurs more quickly in Jewish blood than in Russian blood; tests of race based on this difference proved correct in 91.7% of cases."

Technology, blood, and percentages—all sounding very much like science is supposed to. And presented for the consumption of genetics students without so much as a skeptical glance at the test itself. This only goes to show, however, how easy it is to find what you're looking for. Where the researchers expected to find essentialized differences marked in the blood of different peoples, they easily found them.

All of which returns us to the question of what's really "out there" when we examine the human species biologically.

And the answer was known two centuries ago to Buffon and Blumenbach.

Whether we examine people's bodies or sample their genes, the pattern that we encounter is very concordant. *People are similar to those geographically nearby and different from those far away.* Dividing human populations into a small number of discrete groups results in associations of populations and divisions between populations that are arbitrary, not natural. Africa, for example, is home to tall, thin people in Kenya ("Nilotic"), short people in central Africa ("Pygmies"), and peoples in southern Africa who are sufficiently different from our physical stereotypes of Africans (i.e., *West* Africans) as to have caused an earlier generation to speculate on whether they had some southeast Asian ancestry. As far as we know, all are biologically different, all are indigenously African, and establishing a single category (African/black/Negroid) to encompass them all reflects an arbitrary decision about human diversity, one that is not at all dictated by nature.

Furthermore, grouping the peoples of Africa together as a single entity and dividing them from the peoples of Europe and the Near East (European/white/Caucasoid) imposes an exceedingly unnatural distinction at the boundary between the two groups. In fact, the "African" peoples of Somalia are far more similar to the peoples of, say, Saudi Arabia or Iran—which are relatively *closer* to Somalia—than they are to the Ghanaians on the western side of Africa. And the Iranis and Saudis are themselves more similar to the Somalis than to Norwegians. Thus associating the Ghanaians and Somalis on the one hand (as "Negroids"), and Saudis and Norwegians on the other (as "Caucasoids"), generates an artificial pattern that is contradicted by empirical studies of human biology.

Immigrants to America have come mostly from ports where seafaring vessels in earlier centuries could pick them up—hence our notion of "African" is actually *West* African; and our notion of Asian is actually *East* Asian. When we realize that people originating from very different parts of the world are likely to look very different, and combine that with the fact that most European immigrants came from north-central Europe, it is not hard to see why we might perceive three types of people. If there were a larger immigrant presence in America representing the rest of the world—western Asia, Oceania, eastern or southern Africa, the Arctic—we would be more struck by our inability to classify them easily as representatives of three groups. Perhaps the most obvious example would be South Asians from India and

Pakistan who are darkly complected like Africans, facially resemble Europeans, and live on the continent of Asia!

One occasionally still hears the suggestion that West Africans represent the purest strains of blacks; East Asians the purest strains of yellows; and northern Europeans the purest strains of whites. The people in between would then be hybrids of varying proportions of the original pure strains. But this is actually a pre-Darwinian gloss on the human species. What biohistorical model underlies this suggestion? The answer is simple: it's the familiar biblical story of Noah, the mid-nineteenth-century version of which had Ham emigrating to West Africa, Shem to East Asia, and Japheth to northern Europe, where their descendants were fruitful, multiplied, and commingled at their boundaries.

But this is contrary to both modern data and theory. In the first place, we know why human populations look different from one another—they have adapted to diverse climates and have been subjected to vagaries of the history of their gene pools. And we also know why they look similar to their neighbors—wherever populations coexist, what Cole Porter called "the urge to merge" has invariably expressed itself. Even Neandertals had trade networks; and where goods flow, so do genes (one can imagine traveling salesmen jokes even in the Pleistocene). So the pattern of general geographic continuity we encounter among human populations makes a good deal of sense.

And in the second place, there does not seem ever to have been a time when people lived only in the most remote corners of the world, say Nigeria, Norway, and Korea. People seem to have been living in the in-between places for as long as there have been people.

What the "pure race" model does is to confuse the most extreme peoples with the purest or most primordial. The people of Lagos, Oslo, and Seoul look the most different from one another *not* because they are the most untainted descendants of the original, physically divergent founders of the species but because they live the farthest away from one another and have consequently adapted to the most different environments, while maintaining the least amount of genetic contact.

Which takes us away from the question of fundamental divisions of the human species, and toward that of describing the diverse adaptations of human populations, as Buffon attempted to do in the mid eighteenth century, while resisting the general classificatory innovations of Linnaeus. In anthropology, this conceptual revolution occurred in the years directly following World War II, as the cultural basis of classification generally became more broadly appreciated, and as the destructive results of accepting such classifications as naturally based became more apparent.

Dividing and classifying are cultural acts and represent the imposition of arbitrary decisions upon natural patterns—especially in the erection of boundaries within a species where none exists in nature. This is most evident in the legalities of defining races, so that intermarriage between them could be prohibited—the miscegenation laws.

"Miscegenation" is a term coined in a satirical pamphlet during the Civil War to mean "racial amalgamation," but intended to sound worse. As late as the 1950s, nineteen U.S. states had laws forbidding it on their books. All nineteen forbade black-white marriages, but many of them specified other races as well—for example, "West Indian, Mongolian, Japanese, or Chinese" in Georgia; "Malay" in Maryland; and "Indian" in Virginia and South Carolina. From a practical standpoint, in order to forbid a marriage, you had to define the criteria of inclusion in the excluded groups. Indiana thus prohibited marriages between a white and any person "possessed of one-eighth or more of negro blood" (Stat. §44-104).

Thus, a single black great-grandparent was sufficient to establish a person as "black," while seven white great grandparents were insufficient to establish one as "white." Some states, such as Florida (Const. Art. XVI, §24; Stat. §741.11), took it back another generation; while Georgia (Code §53-106; ch. 53.3) demanded "no trace of [non-white] blood in their veins."

Look carefully. You'll realize that race is being inherited according to a symbolic or folk system of heredity, in contrast to biological inheritance. Thus, racial heredity is qualitative, all-or-nothing—any amount is as good as any other; while biological heredity is quantitative and fractional.

This naturally leads to confusion. One can read an article about the legal scholar Lani Guinier, for example, and learn in one place that she is "black" and in another place that she is "half-black." How can somebody be both black and half-black at the same time? Algebraically that would seem to describe the equation $x = \frac{1}{2} x$, to which the only solution is $x = 0$. Of course, the confusion stems from the fact that, as the product of a "mixed marriage," she is "half-black" *biologically*, and "black" *socially*. Half of the biological contribution is ignored in the construction of a social identity. The biological and social realities contradict each other, and the social reality, based on a folk model of heredity, dominates.

This distinction between fictive or "folk" ideas about heredity, and scientific or biological patterns of heredity, is crucial. On the one hand, we want to find out what is "really out there"; and on the other hand, we impose order upon what is out there by carving it up culturally and imagine that these divisions accurately describe it.

Race turns out to be an optical illusion. It's not the pattern of human differences we encounter empirically. Humans vary principally locally, not continentally.

"CHIMPANZEE" AS A RACIAL INSULT

Recognizing, however, that humans do not come naturally packaged in continental groups, color-coded for your convenience, does not preclude prejudice or discrimination. Historical prejudice in America against Jews and the Irish, and contemporary biases against Latinos, women, homosexuals, and many other classes of people in different places and times, show this point clearly. Prejudice is independent of biology or genetics.

The chimpanzee plays a classic and tragic role in the drama of human hatreds, for it is a creature similar to us, yet an animal; a beast, even if a clever one. Likening people to chimps is therefore a convenient symbolic means of denying their humanity. In the old metaphor of the "great chain of being," chimpanzees are something less than human. By likening the Africans or the Irish to chimps, as both were sometimes caricatured (the Victorian novelist Charles Kingsley, for example, wrote explicitly to his wife of the "white chimpanzees" in Ireland), you deny their essential humanity, and thereby their worth as humans.

More subtly, though, there are properties culturally assigned to the apes—often themselves slanders—that also serve to degrade humans. The most popular cultural stereotypes of the apes, for example, involve their stupidity and licentiousness—stereotypes that were, of course, also applied readily to immigrant, impoverished, or otherwise excluded groups of people.

Thus, Buffon in the mid eighteenth century emphasizes the unity of the human species (in contrast to Linnaeus's subspecific classifications), but remarks casually on "the forced or voluntary matings of Negresses and apes." The apes, after all, were thought to be lustful for human females; and maybe the African women they snagged were reciprocally lustful for ape males as well.

It doesn't necessarily have to be chimpanzees, of course, although they are the most convenient, being the most nearly human of the nonhumans. Consequently, when Frederick Goodwin, then the director of the Alcohol, Drug Abuse, and Mental Health Administration, observed in 1989 that high levels of lethal violence in urban slums *literally* suggested "life in the jungle" to him, although he was invoking rhesus monkeys as his dehumanizing species of choice, everyone knew what he meant.

No one, however, quite knew what to make of a 1996 book on "apes and the origins of human violence" that prominently featured a gorilla and a black man on the cover. Co-authored by a Harvard professor with ostensibly liberal political sensibilities, the book had simply used a classic old illustration for its cover art. The picture in question had been drawn by Adolph Schultz, a Swiss anatomist at Johns Hopkins, who was arguably the leading primate anatomist of his day. Schultz was also a gifted artist and illustrated his scientific papers with magnificent ink drawings. In 1933, Schultz published a figure illustrating the relative body proportions of a chimpanzee, a gorilla, an orangutan, and a human being in the German journal *Anthropologischer Anzeiger*. The subjects are depicted with their bodies shaved, and even with the skeletons partly showing through their skin (fig. 6).

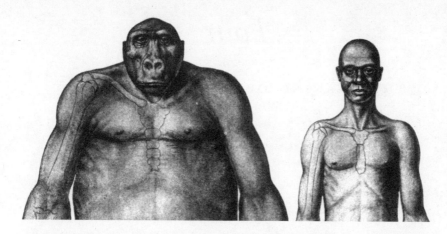

Figure 6. Drawing by the Swiss anatomist Adolph Schultz illustrating the relative body proportions of a gorilla and a man. From A. H. Schultz, "Die Körperproportionen der erwachsenen catarrhinen Primaten, mit spezieller Berücksichtigung der Menschenaffen," *Anthropologischer Anzeiger* 10 (1933): 154–85.

The illustration was widely reprinted, but there was one problem. Schultz's human subject was an excellent representation of a corpse he had dissected, which he identified as "adult Negro." It was simply a depiction of the human specimen he had most readily available, but of course it fed into a number of archaic ideologies, within which the "highest" ape would be contrasted with the "lowest" human. An indignant article in the *American Anthropologist* in 1965 observed that Schultz's drawings commonly compared blacks with apes, suggesting (sometimes explicitly) that whites were less apelike, and called the practice "mischievous and incendiary." Thus, when reprinted by anthropologists, the illustration has often been altered to show the four subjects merely in silhouette.

Consequently, it was a bit jarring to see the original figure not only on the cover of a 1996 book but specifically cropped to emphasize the comparison between the gorilla and the black man.

The Origins of Human Violence, it said.

What could the publishers have been thinking?

Four

THE MEANING OF HUMAN VARIATION

PEOPLES OF THE WORLD DIFFER from one another, and to understand the nature of those differences we are obliged to compare them. The social issues involved in such comparisons, however, necessitate considerably more introspection than would be taken for granted by a scientist accustomed to comparing spiders or earthworms.

Skin, hair, face, and body form all vary across the world's populations. In humans, these biological differences are complemented and exaggerated by differences in language, behavior, dress, and the other components of the cumulative historical stream we call "culture." The skeletal differences among the world's most different peoples are actually quite subtle, however, so that although trained forensic anthropologists can allocate *modern* remains into a small number of categories given to them, it is much harder to sort prehistoric remains, and almost impossible to sort skeletons accurately from places other than West Africa, East Asia, and northern Europe. If you put human bones from the Middle East, North Africa, and New Guinea together in a pile, they would be almost impossible to sort accurately, and it would actually be impossible to tell even how many geographic regions were represented. Why? Because you would no longer be distinguishing among discrete and familiar populations; rather, you would now be obliged to distinguish among connected and unfamiliar populations.

The recognition that indigenous peoples of the world were not naturally packaged into discrete groups was slow to come. It was comparable to the replacement of the more intuitively obvious earth-centered solar system by the sun-centered system. It was a veritable revolution in the way we interpret the world and the way we see ourselves in relation to others.

In anthropology, this conceptual revolution was catalyzed during the years directly after World War II by the anthropologist and author Ashley Montagu, whose articulate liberal humanism struck a responsive chord with both professional and popular audiences; and by Sherwood Washburn, a former pupil of Earnest Hooton's, who forswore the static research program of the previous generations and articulated a view of a "new physical anthropology"—one that would emphasize adaptation and population, not race and classification.

In his mid-nineties, Ashley Montagu was still an imposing figure (he died in late 1999), combining the physical grace of a former boxer and the striking profile of an English aristocrat. But the gene pool of the English aristocracy had nothing to do with it. Born in turn-of-the-century London, he adopted the name Montague Francis Ashley Montagu to gain entrée into circles that would not welcome a person named Israel Ehrenberg. As a youth, he read voraciously and entered anthropology during its heyday at the London School of Economics, under the charismatic professor Bronislaw Malinowski. He augmented this training in cultural anthropology with an apprenticeship to the leading anatomist in England, Sir Arthur Keith. After emigrating to New York in the 1920s, Montagu entered the circle of the equally charismatic anthropologist Franz Boas at Columbia.

In 1941, Montagu launched the first in what would become more than a half-century of attacks upon the formal scientific classification of humans in physical anthropology. This naturally brought him into conflict with the powerful Harvard professor Earnest Hooton, who had established that very problem as the raison d'être of the entire field. In the 1950s, Montagu left academia as a full-time livelihood—actually driven from it by right-wing zealots of the McCarthy years—

to devote himself to popular writing and lecturing, and became one of the best-known public intellectuals in America, second only to Margaret Mead as a publicly known anthropologist. He could be seen, for example, debating the psychologist Richard Herrnstein on *The Phil Donahue Show* on the subject of the genetic determination of intelligence and behavior a full quarter-century before *The Bell Curve.*

Other than their shared interests, Montagu had almost nothing in common with Sherwood Washburn. The two were never particularly close friends. Standing a foot and a half shorter than Montagu, Washburn led a classically academic life (he died in 2000, at the age of 88). He was from old New England stock, the son of a minister, Harvard-educated. Aside from a few articles for *Scientific American* over the span of a few decades, he disdained both popular writing and book-length expositions; Washburn's milieu was the research paper and the inskirts of academia—presidencies of learned societies, editorships of academic journals, and a host of supervised doctoral theses. He preferred the term "biological anthropology" to his mentor Hooton's "physical anthropology," in order to emphasize the dynamic, evolutionary component of the field. Washburn's alienation from Hooton's ideas began in graduate school with his growing interest in the anatomical adaptations of primates, and he began to look at humans through the same lens.

Washburn's first faculty post was at the Columbia University Medical School, where he taught anatomy. Columbia had been the academic center of Franz Boas's anthropological school, and that of his inheritors, Ruth Benedict and Margaret Mead. And it was in New York that Washburn and Montagu independently befriended the brilliant population geneticist Theodosius Dobzhansky, a Russian émigré whose 1937 book, *Genetics and the Origin of Species,* had revolutionized the field. Washburn recalled that at his first meeting with "Doby," the geneticist half-scowled at him, saying: "So, you studied under Hooton? I don't understand the way he divides people into racial types. Explain it." To which the young assistant professor replied: "I can't. I think there's only populations, not racial types." At which point the geneticist circled around to the front of his desk, grabbed Washburn's hand, and shook it vigorously.

Washburn and Dobzhansky organized a major convocation of geneticists and anthropologists in 1950 to discuss the integration of the two fields. Of the eighteen anthropologists who spoke, eleven had studied under Earnest Hooton at Harvard. Montagu was one of the other seven. The meeting's theme was the shift away from studying nebulous abstract "races" and toward the analysis of real populations and their adaptations and differences; and the goal was reiterated time and again. Hooton himself was in attendance, and Washburn recalled his old professor taking him aside during the conference and telling him, "Sherry, I hope I never hear the word 'population' again!"

The professional outputs of Washburn and Montagu suddenly dovetailed in 1962, however. Postwar anthropology had not fully come to grips with prewar anthropology, because Nazi racial anthropology was actually not that different from American racial anthropology. Hooton had consistently ridiculed the German anthropologists' work, but many of the differences were actually subtle, and the criticisms directed at the German scientists could hardly be evaded by American scientists, including Hooton himself. Hooton, after all, freely speculated that the technologically backward cultures of Africa were what comes of inferior brains, and inferior brains were the natural products of inferior genes. Wasn't that pretty much what the Germans thought too?

By the early 1960s, however, Nazi racial anthropology had been appropriately demonized; American racial anthropology was finally disintegrating; and the civil rights movement was crystallizing. But the changes did not come without a struggle.

ON THE ORIGIN OF RACES

Carleton Stevens Coon was a New England Yankee and had been one of Hooton's first students. A great bear of a man, he recounted his exploits in the field with bravado and could drink hard and cuss with the most macho in any party. Coon taught at Harvard and Penn and was recognized for his research and publications by his election as president of the American Association of Physical Anthropologists in 1961. At the time, he was writing his magnum opus, tentatively titled *On the Origin of Races,* to invoke Darwin.

The book was published the following year, having dropped the first word of the title. *The Origin of Races* was both engaging and authoritative. It was also provocative, in a time and place struggling to slough off the premodern science that had kept the nation divided and much of it oppressed. Coon presented four major theses in the book: (1) that the human species was divisible into five races, or subspecies—Caucasoids, Negroids, Mongoloids, Australoids, and Capoids (southern Africans, from the Cape of Good Hope); (2) that these five races were recognizable in the Pleistocene fossil record and had evolved largely independently of one another for the past half-million years; (3) that they had evolved into *Homo sapiens* in a specific sequence—namely, Caucasoids, then Mongoloids, then Negroids, and last, Capoids and Australoids; and (4) that "the length of time a subspecies has been in the sapiens state" was the cause of "the levels of civilization attained by some of its populations."

In other words, Coon sought to explain civilization as a product of nature, not as a product of human agency. And, more important, he sought to explain political and economic dominance in the same way. This of course revisited the most odious of archaic racist ideologies, implying that slavery and oppression were good and necessary, because the enslaved and oppressed were constitutionally inferior to the enslavers and oppressors.

By 1962, however, pretty much all anthropologists recognized that "levels of civilization" were the results of social history, not of biological endowments. Anybody could have built the pyramids, but the Egyptians did. And so did the Mayans. And so did other peoples. But how could you hold the history of all the peoples who did *not* build the pyramids against them? Peoples have their own histories, and it's nobody's fault that Indonesia didn't fight in the Punic Wars, or that Eskimos didn't invent refrigerators. You can't really hold the fact that they didn't have someone else's history against them. Coon's division of the human species was coupled with a scientific scenario for the biological origin of the species, which in turn formed the basis of a theory that purported to explain the social, political, and economic subjugation of the dark-skinned indigenous peoples of the earth by recourse to their biological natures.

And the message was not lost on political activists of the era, in particular the opponents of civil rights and supporters of segregation. As one segregationist propagandist, a sometime airline president and historian named Carleton Putnam, would argue, based on the work of his cousin, Professor Coon: How can you put black and white children together, when the latest scientific research shows that blacks are 200,000 years of evolution behind whites?

Coon's book was reviewed everywhere, from the *New York Times* to the *Harvard Crimson* (by an undergraduate pre-med anthropology major and aspiring novelist who signed his review "J. Michael Crichton"). It was discussed on television and radio, in editorials and in barber shops. It was physical anthropology, the science of human classification, made relevant to the social issues of the day.

Ashley Montagu and Sherwood Washburn thus ended up on the same side, working to purge anthropology once and for all of the classificatory fallacy that had blinded it since the time of Linnaeus. Montagu attacked the book in professional and public forums; Washburn used his bully pulpit as president of the American Anthropological Association to denounce the old, essentialized studies of human variation represented by *The Origin of Races*. And although Washburn made only a single oblique reference to Coon's book, everyone knew what he was talking about (indeed, as a graduate student at Harvard in the 1930s, he had once been assigned as a teaching assistant to Coon's course). Coon died in 1979, an embittered and largely forgotten figure, done in, he supposed, by the forces of political correctness, and more darkly (he allowed in personal correspondence) by a conspiracy of communists and Jews as well.

In fact, Carleton Coon was a tragic anachronism, a throwback to an era and literature whose claims to science and scholarship had long been exposed as pretenses. There is always a purpose for comparing peoples—if you find their feet to be bigger, it explains why they run faster or slower, depending upon whether you are intrigued by how fast or slow they run. If their heads differ, that explains why they are smarter or dumber; if their eyes differ, that is why they see more or less distinctly.

Can we contrast different populations of the world? Of course. But that's not the important question. The important question is the one that lurks underneath: Why do we *want* to contrast different populations? Science begins with the framing of an issue, a reason to collect and examine data. Without that frame, without a reason to collect data, data are meaningless and the endeavor begins and ends as futile pseudoscience.

Populations differ in many ways. In fact, if you compare two groups of people, *any* two groups of people—painters versus chemists, Yale freshmen versus UCLA freshmen, the New York Yankees and the Boston Red Sox—you are very likely to find some differences between them. What makes this endeavor scholarly or scientific—what transforms it from information into knowledge—is the interpretation you give to the results, how you impart meaning to the data. And this is determined by the question you have in mind when you actually do the comparison.

Consider a classic comparison, brain size, across the classic three races—black, yellow, and white—as was carried out in the mid nineteenth century by a premodern physical anthropologist, Samuel George Morton. Assuming there are such discrete groups, and that one can make do with samples from anywhere as representative of those groups, Morton amassed an unparalleled collection of skulls from around the world, stuffed them with mustard seed or lead shot, and measured how much each skull could hold. On average, his fifty-two "Caucasian" skulls held 87 cubic inches of stuffing, or about 1426 cubic centimeters; his ten "Mongolian" skulls, 83 cubic inches, or 1360 cc; and his twenty-nine "Ethiopian" skulls, 78 cubic inches, or 1278 cc.

What hypothesis guided Morton? Why did he bother to do this? To see whether—to show *that*—his intuitive ranking of the intelligence of the races could be also read from their physical properties. And indeed it could—whites were bigger-headed (and smarter); and blacks smaller-headed (and dumber). But what twenty-nine skulls could adequately have represented the diverse peoples of Africa, whom he called "Ethiopians"? Were they actually from Ethiopia, or were they American slaves? (Black Americans have hardly any ancestry from Ethiopia, which is in *East* Africa.) Were the samples all taken from people the same size? Smaller people tend to have smaller heads, after all. Were the samples all taken from people of roughly the same age? Were they taken from people of the same class? Better-nourished people tend to have bigger bodies and heads.

The point is that the study was being undertaken for a reason. Morton chose to measure a certain variable in a sample of people clustered a certain way, and he used the results to demonstrate a certain point. If the point had not been supported, nobody would ever have heard about it. This is not necessarily to suggest any wrong-doing, as Stephen Jay Gould does strongly in his classic *The Mismeasure of Man,* but simply to observe that it is easy to collect data that support your position. And once that position is supported, it is easy not to look too hard for other explanations.

Some 150 years after Morton, Canadian psychologist J. Philippe Rushton again compares crania sorted by race in order to show that brain size is a predictor of "intelligence." And what does he find? In an era in which Asians excel academically, their brains are now bigger than European or African brains! Unsurprisingly, Rushton's interpretations are not the obvious ones—that brain size seems to track the measurer's expectations; or that people's brains mysteriously seem to expand in size as they attain higher educational levels—but rather simply, again, that blacks are inherently dumb.

So these comparisons are far from benign. Comparisons are made to prove a point. And in a racially fixated society, the purpose the physical comparisons are used for is generally going to be unflattering to somebody.

Certainly, the apparent malleability in brain size over the past century, with Asians craniometrically leapfrogging over Europeans, directly implies that (1) Africans could leapfrog over both; and (2) these physical measurements are too crude to be biologically, anthropologically, or socially meaningful. We will return to the brain in the next chapter; for present purposes, we may simply note a survey by Colin

Groves, an anthropologist at the Australian National University. Recognizing that human biological variation is local, not continental, Groves looked at skulls that way—and he found exceedingly large brains among some aboriginal populations of Hawaii, Mongolia, France, South Africa, and Tierra del Fuego; middling brains among some aboriginal inhabitants of Korea, Norway, Hawaii, and East Africa; and small brains among some populations of Sicily, Peru, India, and West Africa. Indeed, skulls do vary, and they do so like everything else—locally. The causes of such variation are complex, but there seems no obvious justification for grouping them into a "racial average." If this tells us anything at all about the relative brains of the races, it must be information of an exceedingly occult nature, except perhaps to someone with an invidious motive.

And that is the problem with comparative racial studies. It is easy to get misinformation, or poor interpretation of data, into the record, and hard to get it back out again. Racial comparisons are invariably shot through with culturally loaded assumptions, and once populations become the level of analysis, the racial generalizations quickly break down.

Consider another classic issue of racial comparisons: are blacks better athletes than whites? We will explore other aspects of this loaded question later, but here is one way in which it has been approached scientifically. A group of Scandinavian marathon runners was compared to a group of Kenyan marathon runners and surveyed for enzyme concentrations and other biochemical minutiae.

Differences were found.

Of course differences were found. But are they *racial* differences? Are they populational differences? Are the differences due to the divergent life histories of the Kenyans and the Swedes? And do these differences have anything at all to do with black-white differences in American athletics, given that American whites generally do not have ancestry from Scandinavia and American blacks generally do not have ancestry from East Africa? To give racial significance to the result involves rooting it in the symbolic representation of Swedish runners as "whites" and Kenyan runners as "blacks." It has little to do with the empirical nature of the results.

And the same situation obtains with genetic data.

One of America's leading population geneticists is Richard Lewontin of Harvard, a former student of Dobzhansky's at Columbia. In 1972, Lewontin took the genetic data that had by then been assembled on the diverse populations of the world and subjected them to a new statistical analysis. What he wanted to know was: Just how much of the detectable genetic variation in our species is to be found in the categories of race? In other words, if you analyze our species as (1) individuals versus individuals within a population; (2) populations versus populations within a "race"; and (3) "races" versus "races"; what proportion of our diversity is encompassed by the third category?

Lewontin's conclusion went against the popular ideology of race, and for that reason was of the utmost significance. And yet it makes perfect sense. The very first genetic system studied, the ABO blood group, was thought to reveal races, because the scientists believed that races were genetically significant divisions, and this was genetic data. But it reveals no such divisions. It reveals two things: first, as we have seen, people intergrade into their neighbors; and second, all populations have nearly all the variation. In other words, except for very rare local mutants and their relatives, a person with a given genetic variant might come from anywhere. Someone with type A blood could be from Africa, Asia, or Europe. So could someone with type B blood. Or someone with type O. Or type AB. The only thing that varies is the probability: the B gene is more common in Asia, so someone with type B or AB blood is slightly more likely to have some Asian than African ancestry; but it's nothing to bet the farm on.

The same is true of the Rh blood-group gene. People can be Rh-positive or Rh-negative. And you find both kinds of people everywhere. This is a ubiquitous polymorphism—genetic variation that exists all over the world. Somebody with Rh-negative blood is more likely to be a Basque than to be a Cherokee (since the Basques have famously high proportions of Rh-negative blood), but is most likely to be neither. In other words, a global cataclysm that killed or sterilized every human being except the Uzbeks or the Dinka would still leave the vast majority of the major genetic variants of the species represented in the gene pool. This is true for Rh, for ABO, and for nearly all human genes.

And that was Lewontin's discovery. The overwhelming bulk of detectable genetic variation in the human species is between individuals in the same population. About 85% of it, in fact. Another 9% of the detectable variation is between populations assigned to the same "race"; while "interracial" differences constitute only about 6% of the genetic variation in the human species.

The implication of the new genetic analysis is that so-called racial variation is a tiny issue to someone interested in the study of general human variation. Its components, its patterns, are almost negligible racially. Race is thus a very small, and therefore trivial, corner of the picture. The argument could of course be made that the 6% of genetic variation that is racial is necessarily the most important 6%. But that would be idle speculation or simple prejudice. Racial variation has now been shown to be scientifically, mathematically trivial.

It does need to be recalled here that population geneticists are nevertheless still practicing the old "bait-and-switch," analyzing data on variation in blood-group genes and enzymes and not analyzing data on variation in genes for skin color, hair form, or nose shape. The fact is that we do not have any data on those genes, nor any information on where they are, how many there are, what they actually do physiologically, or how they differ. The assumption is that the genes we can study follow the same pattern as the genes we are interested in.

And there is every reason to think they do. All populations have people who are taller and shorter, fatter and thinner, braver and more timid, introverted and extroverted, more decisive and more dithering. There is no continent of uniformly brunet people to contrast with one of uniformly blond people, or of uniformly tall people to contrast with one of uniformly short people. The traits that differentiate Swedes and Kenyans at the "racial" level are actually a very small proportion of the differences detectable, which are distributed primarily among members of the same population and between populations on the same continent.

A study that captured the public's imagination was the "Mitochondrial Eve" report published at the beginning of 1987 by Rebecca Cann, Mark Stoneking, and Allan Wilson at Berkeley. Mitochondria are, as noted earlier, "the powerhouses of the cell," containing their own 16,500 bases of DNA, distinct from the 3.2 billion bases of the chromosomes. Moreover, at fertilization, a human egg retains the maternal cellular mitochondria, and sloughs off the sperm's mitochondria. This DNA is thus inherited directly through the mother's lineage and is not subject to the vagaries of Mendelian assortment and recombination. A child is a mitochondrial clone of its mother and unrelated to its father.

For Cann's doctoral thesis, she undertook the first direct survey of nucleotide variation in the human species. She sampled a restricted range of bases of the mitochondrial genome (approximately 10% of it) across nearly 150 people, drawn from diverse populations. She then took the resulting diversity and asked her computer to produce a tree linking the various individuals together by virtue of the similarity of their mitochondrial genomes (or mtDNA—sometimes rendered as "empty DNA" by skeptics). The tree was constructed from the 140-odd individuals studied, not from any a priori groupings. When the geographic origin of the samples was imposed on the result, it was found that Africa (represented by African Americans, on the assumption that any admixture was from white males and therefore not passed on in mitochondria) constituted the most genetically diverse continent. The origin of variation in mitochondrial DNA diversity was thus located in Africa, and since mtDNA is passed on only through the mother's lineage, the mtDNA of interest would necessarily have belonged to a female. This African progenitrix was given the biblically inspired name "Mitochondrial Eve."

Most aspects of the study and of the evolutionary model that accompanied it have been effectively criticized in the past decade, but the basic findings seem solid: that racial clusters are not inherent in genetic comparisons of humans but must be imposed by the investigator; that Africans are more genetically diverse than Europeans and Asians, and that they indeed subsume the genetic diversity found in the rest of the world. Why should this be the case? Presumably because modern humans evolved in Africa and some only subsequently emigrated to the rest of the world.

This carries a number of implications for the study of the human gene pool. It means, most significantly, that Africans must be considered paraphyletic; that Africans are an ancestral diverse group and some Africans are more closely related to some Europeans and Asians than to other Africans. Africans, Europeans, and Asians therefore do not constitute comparable groups, even though they may make their homes on different continents. Rather, although Europeans and Asians have, of course, evolved some of their own specializations over the millennia, they are genetically subsumed by the gene pool of Africans.

Comparing Europeans, Asians, and Africans is thus something like comparing dogs, cows, and mammals. Such a comparison is meaningless, because the third category incorporates the first two.

Comparable data were slow to arrive for the DNA of the nucleus, because it evolves more slowly (and thus there aren't many differences among people to tabulate), and because nuclear DNA comes in two different copies per person, as opposed to the uniformity of the mtDNA sequence in a single individual. A few years later, however, in 1996, another young female graduate researcher, Sarah Tishkoff, devised a clever DNA comparison of her own at Yale.

Tishkoff (now at the University of Maryland) took advantage of the major difference between mitochondrial and nuclear DNA, which is usually perceived as a liability in studying nuclear DNA—namely, that different variants of the same gene occasionally trade positions with one another. In other words, if one chromosome 8 in your cells has genes A and B, and the other chromosome 8 has genes a and b, most of your reproductive cells will have AB and ab, as you do, but a small percentage will have gene A and gene b together, and reciprocally the other chromosome will have aB. This is known as "recombination," and it permits—indeed, mandates—linked combinations of genes to be broken up and brought into new combinations.

At the level of populations, and over the span of evolutionary time, all genetic variants ultimately recombine and thus can be found in all possible combinations. But this requires enough time for these recombination events to occur. Recombinations are individually rare, but summed over all reproductive cells for all people over many generations, they always happen. Thus, particular genetic variations that are always found together simply haven't had enough time to come apart.

So Tishkoff chose to look at the linkage relationships of two genes over the world's populations. The first gene came in two forms; we'll call them A1 and A2. The second came in twelve different forms, B1 through B12. She found the fullest array of diversity in her ten sub-Saharan populations; A1 was found coupled with B1 through 12 in someone or another, and A2 was found coupled with all the Bs except B1 and B7, in varying frequencies. By contrast, in her European sample of seven populations, she found A1 joined to B2 and B7 frequently, and to B3, B4, B6, B8, and B9 rarely; and A2 joined to B3 frequently, and rarely to B2, B4, and B7. The rest of the combinations didn't exist at all in Europeans.

Tishkoff's interpretation was properly the simplest. The populations of sub-Saharan Africa were the oldest; eons of recombination had yielded many different combinations of these genes. But a small group of emigrants out of Africa had carried a restricted set of these genetic combinations. And they had not had sufficient time to regenerate the full scope of variation by recombination.

As if that were not enough, the three populations she examined from the northern and eastern parts of Africa had patterns and levels of diversity intermediate between the high sub-Saharan African and low EurAsian levels. Thus the nuclear DNA survey parallels the findings of mitochondrial DNA.

Another significant conclusion of the genetic research study bears noting as well. How genetically different are two random humans from each other? Being good primates, we tend to put a lot of significance on obvious, visible features of the face. But that, as Lewontin's study suggests, might not reflect much of the underlying genome; we could be facially different as a result of very few genetic differences and genetically nearly identical otherwise.

If we look to our closest relatives, the chimpanzees and gorillas, we find an odd pattern. Two chimpanzees or two gorillas are considerably more different from each other than two humans are. In other words, even though chimps, gorillas, and humans diverged from one another about seven million years ago and are all consequently the same age, humans appear to be depauperate in genetic variation compared to our closest relatives.

Lanky, blue-eyed Amos Deinard was breaking hearts in Yale's anthropology department from the first day he arrived from Minnesota. But he only had eyes for chimpanzees and yearned to combine a definitive molecular anthropology survey with training in veterinary care. He had ambitions to be expert in all the major aspects of chimpanzee biology, at both the organismal and population levels, starting with their genes. When he finally finished his doctoral thesis, I said to him, "Goodbye, Mr. Chimps."

Allan Wilson's group at Berkeley, pioneering the analysis of mitochondrial DNA, had shown that the apes were considerably more diverse than humans. But the strange rates and modes of mitochondrial DNA evolution left unclear whether that pattern would also obtain for nuclear DNA, the real stuff. That is what Deinard wanted to discover. The advantage of studying mitochondrial DNA was that it was easy to work with, and each individual only had one sequence; surveying nuclear DNA—even a minuscule sample of it—would be far more painstaking.

He studied five short DNA sequences on different chromosomes, across forty-five chimpanzees, eighteen bonobos, and fifteen gorillas, and compared them to a sample of nearly one hundred people drawn from all over the world. And when the cells had been cultivated, the DNA had been extracted, and the samples had been run, he found a highly consistent pattern. Where the large global sample of humans revealed one or a few DNA sequence variations, the corresponding sequence from these endangered primates of equatorial Africa almost invariably had more. For example, from a region of gene called Apolipoprotein B, located on chromosome 2, Deinard found four DNA sequence variants in his survey of the human species.

By contrast, he found three DNA sequences from his mere eighteen samples of the rare and endangered bonobo. Moreover, he found seven different DNA sequences in his gorilla DNA, and no fewer than seventeen different DNA sequences in his common chimpanzee sample.

Not only did the apes have more sequence variations, but they had more diverse variations—the sequences they had were often more different from each other than two different sequences in humans. And this in spite of the fact that the human sample was much larger than the ape sample and the apes all derived from relict populations in a small part of the world.

Genetic evidence thus points to a recent origin for the genetic diversity within our species. Genetic diversity in the human species is surprisingly ephemeral—only on the scale of tens of thousands of years. How, then, are humans able to make sense of their place in the social world, to know what groups they belong to, and who else does, if they don't have much biological diversity to guide them, as other animals do?

The answer is that genetic variation between groups seems, in some large measure, to have been "replaced" by cultural diversity.

Cultural diversity is the true evolutionary hallmark of humans, and it serves the purpose (among many others) of telling us who we are, of identifying ourselves in the social universe. Irish Catholics and Irish Protestants are indistinguishable genetically, but they know who they are, and who they're not, by virtue of their cultural differences. Any place on earth where there is civil unrest almost inevitably involves people who are genetically and physically very similar and culturally different. The fascinating thing about cultural differences is that they are more important to us, for example, in creating life-and-death situations than biological variation is. Where there are biological differences between hostile populations, they can easily be recruited to mark the enemy. But group hostilities inevitably lie in the social, political, and economic realms, not in the biological.

Cultural variations are very stable (they persist for longer than individual lifespans) and very structuring (the knowledge they impart seems so obvious that it is difficult to imagine ever calling it into question). Yet this was precisely the contribution of anthropology in the twentieth century—to show that the way we often think the world is may not be the way the world comes to us naturally at all. It may be an artifact of the cultural glasses we wear. The fact that populations seem so different from one another—in the way they dress, speak, act, think, eat, move, and live—tempts us to think that those differences must have some sort of basis in their natures, their core beings, their genomes.

Which brings us to the next question.

ARE CONSISTENTLY DETECTABLE DIFFERENCES BETWEEN HUMAN POPULATIONS GENETIC?

If I study 1000 Ibos from Nigeria and 1000 Danes from Denmark, I can observe any number of differences between the two groups. One group, for example, is darkly complected; the other is lightly complected. This difference would probably be the same whether I selected my sample in the year 1900, 2000, or 2100, and is presumably genetic in etiology.

On the other hand, one group speaks Ibo and the other speaks Danish. That difference would also be there if I selected my sample in 1900, 2000, or 2100, but is presumably *not* genetic. At least, generations of immigrants attest to the unlikelihood of a genetic component to it.

How, then, can we know from the observation of a difference whether it is biologically based or not?

European explorers were well aware that the people who looked the most different from them also acted the most differently. Linnaeus invoked broad suites of personality ("impassive, lazy") and culture traits ("wears loose-fitting clothes") in his diagnosis of four geographical subspecies of humans in 1758. The next generation of researchers recognized that these traits were both overgeneralized (if not outright slanderous) and exceedingly malleable, and sought to establish their formal divisions of the human species solely on biological criteria.

It was widely assumed by the middle of the nineteenth century that regardless of the degree of malleability of mental or behavioral traits of human groups, the features of the *body* were fundamentally immutable. Anthropology is, in its fundamentals, about getting inside people's heads. But lacking a way to look literally in people's heads, the next best thing was to look *at* people's heads. And like the rest of the body, heads varied in size and shape.

In 1847, a Swedish anatomist named Anders Retzius devised a way of measuring and comparing heads. He measured the head at its widest point and divided that by a measurement of the head at its longest point, and then he multiplied by 100. People with wide heads would thus come out with a low number, say around 75; and people with long heads would come out with a high number, say around 85. And a highfalutin vocabulary came to match the data: low numbers were called brachycephalic, and high numbers were dolichocephalic. Various other polysyllabic terms quickly entered the craniological lexicon.

This has been satirized in many places, for example, by Sir Arthur Conan Doyle. Sherlock Holmes is introduced to the mystery of the hound of the Baskervilles by a certain Dr. James Mortimer, who exclaims upon meeting the detective: "I had hardly expected so dolichocephalic a skull or such well-marked supra-orbital development." Then he adds pseudoerotically, "Would you have any objection to my running my finger along your parietal fissure?" Somewhat later, in the movie version of the musical *On the Town* (1949), Ann Miller plays a tap-dancing anthropologist infatuated by a sailor's "sub-super-dolichocephalic skull"—whatever that means.

The important thing about skull shape was that it appeared to be a fairly stable feature of people and of populations. It was also studiable in ancient skeletons, so comparisons could be made between archaic and modern populations. And, of course, it measured the head, which encased the brain, which determined the mind, which was composed of diverse thoughts. It *had* to be important. Thus, traits like the shape of the head could be taken as an indicator of transcendent biological affinity—groups with similarly shaped heads were closely related, whereas those with differently shaped heads were more distantly related.

The first to challenge this assumption empirically was the pioneering American anthropologist Franz Boas, who measured skulls of immigrants to Ellis Island, and compared them to those of relatives already living in the United States. Boas was a German-Jewish immigrant who had come to anthropology via a circuitous intellectual route, through physics and geography. With a guiding interest in recording the lifeways of the indigenous peoples of the New World as they were lost or irreversibly altered, Boas came to appreciate the mammoth effects that cultural changes had on people in all spheres—social, mental, and physical. As a Jew in nineteenth-century Germany, he also had firsthand experience with the manner in which socially marked groups can come to have their identities or differences inscribed or etched into their makeup and their second-class status assigned to nature.

Physical anthropologists already knew that the skull was somewhat malleable. Various cultures of the Middle East and the U.S. Southwest used cradle-boards to carry their babies, which grew up with the backs of their heads flattened. The Incas molded their children's skulls to grow high; the "Flathead" Indians of Canada molded theirs to grow low. Might more subtle, long-term influences on the skull affect its shape as well? If so, that would strongly undermine the assumption that skull shape tells you much about race, or population relationships, much less about the inner mind of the person attached to the skull.

European immigrants to America thus constituted a natural experiment for the theory that consistent bodily differences are innate. Keeping the gene pool constant and altering the environment, Boas found that the human body is indeed very sensitive to the conditions of growth. The characteristically round-headed Jews coming to New York from eastern Europe were considerably more round-headed than their relatives already living in New York. And the children of immigrants, the generation born in the United States, were even less characteristically round-headed than their parents. By the same token, the characteristically long-headed Sicilians became less long-headed in America. There was a decided tendency of diverse immigrant groups to become more physically homogeneous in America—in spite of marrying within their own groups—than they were when they arrived.

In particular, the shape of the head turned out to be very plastic and not at all a reliable indicator of genetics, or race. It wasn't that people were turning into one another, but simply that the differing conditions of life tended to exaggerate physical differences among human populations. In similar surroundings, people became more physically similar.

Subsequent studies of other immigrant groups, notably the study of Japanese immigrants to Hawaii by Harry Shapiro and Frederick Hulse in the 1930s, supported this discovery. Immigrants to Hawaii, they found, differed from their relatives in Japan; and the Hawaiian-born generation differed even more markedly. Thus, the observation of consistent difference between groups of people—even of the body—is not necessarily indicative of a genetic basis for that difference. This work effectively shifted the burden of proof from those who question a genetic basis for the observation of difference to those who assert it.

To establish a genetic basis for an observed difference between two populations, therefore, requires more than just observing the difference to be consistent. It requires presumably genetic data. The inference of a genetic difference in the absence of genetic data thus represents not a scientific theory of heredity, but a folk theory of heredity. To the extent that behavioral and mental traits—such as test scores and athletic performances—are even more developmentally plastic than are strictly physical traits, the same injunction must hold even more strongly for them. Genetic inferences require genetic data.

DO DIFFERENT PEOPLES HAVE DIFFERENT POTENTIALS?

One of the catchphrases of 1994's best-selling book *The Bell Curve* (about how welfare should be abolished because the people on it are just plain stupid and there's nothing anyone can do about it) was "cognitive ability." Eluding a scientifically rigorous definition, the phrase is left to be explained by a commonsense or folk definition— "cognitive ability" presumably means the mental development possible for a person under optimal circumstances. But it would take an extraordinarily naïve or evil scientist to suggest seriously that such circumstances are in fact broadly optimized across social groups in our society. Consequently, not only can we not establish *that* abilities are different, we have no reliable way even to measure such an innate property in the first place. What we have is performance—on tests or just in life—which is measurable, but which is the result of many things, only one of which is unmeasurable innate ability.

On the one hand, it is not at all unreasonable to suggest that different people have different individual "gifts"—we all possess unique genetic constellations, after all. On the other hand, those "gifts" are not amenable to scientific study, for they are only detectable by virtue of having been developed or cultivated. Thus no scientific statements can be responsibly made about such genetic "gifts" in the absence of the life history of the person to whom they belong.

In other words, ability is a concept that is generally easy to see only in the past tense. I know I had the ability to be a college professor, because I *am* one; but how can I know in any scientifically valid sense whether I *could have been* a major-league third baseman? I can't, so it is simply vain for me to speculate on it. A life is lived but once, and what it could have been—while fascinating to contemplate—is not a scientific issue.

There is also an important asymmetry about the concept of ability. A good performance indicates a good ability; but a poor performance need not indicate poor ability. As noted above, many factors go into a performance, only one of which is ability. Thus, when we encounter the question of whether poor performance—even over the long term—is an indication of the lack of cognitive ability, the only defensible position from the standpoint of biology is agnosticism. We do not know whether humans or human groups differ in their potentials in any significant way. More than that, we *cannot* know, and that is a crucial distinction. There is no experiment you can devise that will distinguish what one normal child at birth could optimally accomplish as opposed to another normal child. Once again, it may sound like genetics, but it isn't genetics.

Science is not about what is known, but about what is knowable. That is why atoms are within the domain of science and angels are not. It is also why individual potentials, like angels, lie outside the domain of scientific discourse and within the domain of "folk knowledge."

Folk knowledge is not necessarily bad. It is simply different from scientific knowledge. It is also not necessarily wrong. Folk knowledge is the commonsense comprehension of things and processes that science often challenges. But science makes a crucial distinction between the knowable and unknowable, while folk knowledge does not. It doesn't need to. The realm of the unknowable is often much more interesting than the realm of the knowable. How many angels, after all, really could dance on the head of a pin? Will people of the future have brains the size of Volvos? Could Muhammad Ali really have beaten Rocky Marciano? If they had really, really, *really* tried, could Beethoven have played for the Chicago Bulls and Michael Jordan have written the Ninth?

Think about it. Over a drink, perhaps.

The important thing to bear in mind, though, is that it's not science, and it's definitely not genetics. It's a folk gloss on issues about "natures" and "innate constitutions" and "potentials"—but these are all ideas that long predate the science of genetics. They have an independent history. They actually have nothing to do with the science of genetics.

Furthermore, this raises a darker question: What are we to make of scientists who assert the existence of real constitutional differences in ability? If we cannot gauge differences in ability in any reliable manner, if ability is not a scientific concept, it is a corruption of science to assert in its name that one group indeed has less ability than another. From the mouth or pen of a politician, the assertion might reflect ignorance or demagoguery; from that of a scientist, it reflects incompetence or irresponsibility. Scientists are subject to the cultural values of their time, place, and class, and historically have found it difficult to disentangle those values from their pronouncements as scientists. We now recognize the need to define the boundaries of science in order to distinguish the authoritative voice of scientists speaking as scientists from the voice of scientists speaking as citizens. This distinction is vital to keeping science from being tarnished by those few scientists who have chosen to invoke it as a validation of odious social and political doctrines.

Racial classifications represent a form of folk heredity, wherein subjects are compelled to identify with one of a small number of designated human groups. Where parents are members of different designated groups, offspring are generally expected to choose one, in defiance of their biological relationships.

Differing patterns of migration, and the intermixture that accompanies increasing urbanization, are ultimately proving the biological uselessness of racial classifications. Identification with a group is probably a fundamental feature of human existence. Such groups, however, are genetically fluid, and to the extent that they may sometimes reflect biological populations, they are defined locally. Races do not reflect large fundamental biological divisions of the human species, for the species does not, and probably never has, come packaged that way.

Merely calling racial issues "racial" may serve to load the discussion with reified patterns of biological variation, and to focus on biology rather than on the social inequities at the heart of the problem. Racism is most fundamentally the assessment of individual worth on the basis of real or imputed group characteristics. Its evil lies in the denial of people's right to be judged as individuals rather than as group members, and in the truncation of opportunities or rights on that basis. But this is true of other "isms"—sexism, anti-Semitism, and prejudices against other groups—and points toward the most important conclusion about human biology: racial problems are not racial. If biologically diverse peoples had no biological differences, but were marked simply on the basis of language, religion, or behavior, the same problems would still exist. How do we know this? Because they *do* exist, for other groups. The problems of race are social problems, not biological ones; and the focus on race (i.e., seemingly discontinuous biogeographic variation) is therefore a deflection away from the real issues.

A society in which individual talents can be cultivated without regard to group affiliations, social rank, or other a priori judgments will be a successful one—acknowledging biological heterogeneity while developing the diverse individual gifts of its citizenry.

"NEANDERTAL MAN": ARCANE YET RELEVANT

It would seem as though what to call Neandertals is a question of strikingly limited relevance to modern life. Neandertals were those stout, football-headed, muscular people who inhabited Europe and the Mediterranean for a couple of hundred thousand years during the Ice Ages and died out about 32,000 years ago. They used to be spelled with an *h* but aren't any more.

In the paradox of sameness/otherness, the Neandertals have been an enigma for over a century. They are diagnosably different from anatomically modern humans, and yet you cannot get any closer to us. They didn't develop art, but then neither did modern humans for a hundred thousand years. We don't know whether or how well they could speak, or conceptualize their world. They buried their dead in stylized positions and borrowed technologies from their neighbors. We know they were smart and successful, because they were around for a long time, and survived under harsh conditions. An expert can differentiate their shoulder blades or thighbones; a college freshman can easily learn to differentiate a Neandertal jaw and skull from those of modern peoples. But what does that mean? Were they another species of people, or even closer than that—another subspecies?

For about half a century, anthropologists have been content to call them a different subspecies, *Homo sapiens neanderthalensis*. That implies that they were geographically and anatomically distinct, but would not have been biologically incapable of breeding with modern humans.

Of course, we cannot know for sure. There is no class of data we could collect, short of finding a living Neandertal and tempting it with a modern mate, which would tell us whether it could in fact have interbred with us. We just don't know how reproductive incompatibilities evolve—sometimes genetically similar animals can't interbreed, and sometimes genetically different ones can. Often, of course, there is some degree of correspondence between anatomical difference, genetical difference, and reproductive capacities—but it cannot be taken for granted.

Calling the Neandertals a subspecies of *Homo sapiens* is a judgment call about their level of sameness and otherness. But it has a salubrious side effect. If Neandertals are a subspecies of *Homo sapiens,* in contrast to all modern humans, then the question of the subspecies of *living* humans is rendered moot. The question that had guided physical anthropology since Linnaeus—What are the divisions of modern people?—now has no possible answer, for the simple reason that subspecies are as low as you can go in the formal taxonomic hierarchy. For us to recognize all living people as one subspecies and juxtapose them at the subspecies level against extinct forms means that we cannot formally name subgroups of living humans, because there is simply nothing left to call them.

That's good. If there's one thing we don't need, it's a free taxonomic niche that would permit the social differences that envelop us to be formalized biologically once again. We have enough difficulty equalizing opportunities among blacks and whites; we don't need new ways for ideologues to divide us and make it look as though it is science that is actually doing it.

But in the paleoanthropology community, a debate has resurfaced in the past decade that has deep ramifications of an unscientific nature: Should we give Neandertals their own species?

This would, of course, be a judgment call highlighting the "otherness" of Neandertals. They were, after all, different. Maybe they were different enough to be called a species. Among the evidence adduced in support of this was the sequencing in 1997 of a short stretch of Neandertal mitochondrial DNA, amounting to 379 bases, which differed somewhat from those of modern humans; where modern humans were found to vary from one another by up to twenty-four differences, the Neandertal sequence averaged twenty-seven differences, depending upon which human sequence you compared it to. The authors then calculated that the DNA sequences of Neandertals and modern people began to diverge about 600,000 years ago, which suggested to them a different species.

Or maybe not, given that there's no rule of equivalence between divergence of DNA and specieshood. Still a judgment call. It's not exactly a scientific datum, it's a parascientific one, which borders on science but is actually in the realm of the unknowable. Could an extinct form of near-human have interbred with us? Not only *don't* we know, but we *cannot* know. Things we cannot know are outside the domain of science. Of course, those are often the most interesting issues to speculate about.

Here's the problem. If we elevate the Neandertals to the species level, we create a vacuum at the subspecies level, one that our folk ideologies about races and racial differences will all too readily fill.

Is it worth it? Is the judgment call on the taxonomic status of Neandertals, an esoteric problem at best, worth the price of creating a scientific loophole for overstating and naturalizing the social divisions in our world?

It is not as if we have positive knowledge about the species status of Neandertals and would therefore be compromising our scientific integrity in striving to be "politically correct." It is, rather, that there are drastic social consequences to this parascientific judgment call about the reproductive capabilities of fossils, which need to be weighed carefully. The status of Neandertals is not an isolated paleontological problem, nor is it a soluble scientific problem; it is a subjective assessment that reverberates into our formal recognition of modern people as well. And it would easily seem to harmonize with archaic, folk ideologies about the living members of our own species, making it easy for narrowly trained scientists to bring back the "Caucasoids" and "Negroids" as if they had validity beyond the colloquial usages.

If you think that this scenario is far-fetched, consider the pronouncements by a couple of geneticists in a major journal, *Molecular Biology and Evolution,* in 1993: "[H]uman populations can be subdivided into five major groups: (A) negroid . . . , (B) caucasoid . . . , (C) mongoloid . . . , (D) Amerindian . . . , and (E) australoid. . . .

(There are intermediate populations, which are apparently products of gene admixture of these major groups, but they are ignored here.)"

How many fallacies can you find? Not only are these geneticists taking archaic folk ideologies as modern anthropology, but they believe in the "Noah's Sons" model of racial origins, with hybridization occurring in the middle of the world. They are at the cutting edge of molecular studies, and back in the nineteenth century anthropologically. They seem neither to have known nor to have cared what the actual data on human variation, and contemporary understandings, were—and neither did the reviewers of their paper. The geneticists actually found five subdivisions because they sampled people from five distinct parts of the world, and not other places: like many studies before it, its conclusion was simply loaded into their assumptions. It's on account of science like this that it's probably a good idea to keep Neandertals safely at the subspecies level.

The history of racial science bears witness to the fact that no class of data on human diversity is value-neutral. It all comes bundled with cultural ideas. On the one hand, that is what makes this area so interesting to study and to examine critically. On the other hand, it exposes a simple lie—that data and theories of human diversity and human origins can exist outside a cultural context, external to a matrix of values and judgments about others, without consequences and attendant responsibilities.

The DNA evidence itself, curiously, did not last long. Although the data were unimpeachable (and two more Neandertal DNA sequences now give concordant results), the interpretation wasn't. Certainly the Neandertal DNA lay outside the range of human variation, as the Neandertal bodies do. But could it really be shown that the gene pools of modern and Neandertal people diverged 600,000 years ago? This was, it turned out, the same class of data (a particularly rapidly evolving part of the mitochondrial DNA molecule known as the "control region") that had suggested to other researchers a few months earlier that the dog had been domesticated 120,000 years ago—in spite of no evidence for it until 12,000 years ago. Could it be that the problem wasn't Neandertal Man's genes at all, however many specimens are sequenced, but the way the dates were being calibrated?

Yes, it could. A few weeks after the initial report, it was discovered that mitochondrial DNA accumulates mutations up to twenty times faster than had previously been thought. That means either that a lot more change should accumulate in the same amount of time or that the same amount of change should accumulate in a lot less time.

If the mutation rate used to calculate the divergence date was too low, then the date of divergence obtained was too distant—more changes were occurring in less time than the calculations allowed. Just how far off the numbers were is unclear. But given the tabulated amounts of difference in the 379 bases of Neandertal DNA, it now appears that 600,000 years of divergence may well have been a considerably inflated estimate. Along with the 120,000-year estimate for the domestication of the dog. And as for the species status of Neandertals following from their DNA divergence, it is no greater than the divergence of subspecies of chimpanzees from one another.

Anthropology may have got it right after all.

Five

BEHAVIORAL GENETICS

WE KNOW INTUITIVELY THAT nature and nurture somehow combine to make us what we are; we don't need science to tell us that. But science brings authority to the table. Many teachers unfortunately skim over the material because it is "emotional," or they attempt to "present both sides"—usually uncritically, as if genetics and behavior were simply a case of "you say potay-to, I say potah-to." Actually, though, an anthropological approach complements and illuminates the genetic data quite well and shows how much we actually do know about the "nature-nurture debate."

Contemporary genetics relies on the ability to relate an observed characteristic (a phenotype) to a hereditary state (a genotype) causally. In other words, genetics is about the mechanism by which a series of DNA instructions comes physiologically to produce an organism with a particular suite of features.

Some connections are straightforward: if you have inherited the genes for cystic fibrosis, you get cystic fibrosis. There is some considerable variation in the expression of the disease, but it is physiologically determined by the genes.

Other connections are more obscure. Light-skinned people tend to have light-skinned children. Tall people tend to have tall children.

Rich people tend to have rich children. Syphilitics tend to have syphilitic children. How do we make genetic sense of this?

There is an easy answer. Do classical genetic experiments. Isolate the individuals with the properties of interest. Mate them carefully to specific partners. Make sure they have plenty of offspring. Study the offspring carefully, mate them to specific partners, and study the distribution of the traits of their children. Under the appropriately controlled conditions, you will be able to distinguish which traits are genetic and even how many genes are involved.

It worked with fruit flies.

Of course, fruit flies have a decided advantage over humans as genetic subjects. In the first place, you can mate them to each other with godlike impunity. They don't seem to care much. And if they did care, it wouldn't matter. It's *your* experiment. They don't whine, and they have no rights.

In the second place, they breed like, well, flies. Having lots of offspring is a very favorable trait for a genetic subject to have, since genetic inferences are principally statistical. Mendel, after all, began by counting up the hundreds of pea plants with the features he was interested in. If his plants had had only three offspring apiece, he wouldn't have gotten very far.

In the third place, fruit fly generations are only a couple of weeks long. You can study many generations during the term of a single research grant.

And that's why fruit fly genetics is so far ahead of human genetics.

One solution is to analogize directly from fruit fly genetics to human genetics. After all, we're both animals and share a basic biology. But just how basic? A fruit fly and a human male both have an XY chromosome configuration; the females have XX. That would suggest that sex determination is similar in both species. But it isn't, actually. The similarity is superficial, for an XXY human is an abnormal male (with a constellation of features comprising "Kleinfelter's Syndrome"), while an XXY fruit fly is a less abnormal female. That suggests instead that something very different is going on in the two species.

Studying behavior adds a further difficulty, namely, identifying the trait in question. Physical traits are there to be studied, measured, analyzed. Behavioral traits must be elicited. And even then, it is easier to analogize to human behavior superficially than to prove a meaningful biological connection.

In a classic midcentury study by A. J. Bateman, fruit flies were carefully mated to a series of members of the opposite sex. The question was, does a succession of lovers increase your reproductive output? Having three partners enabled a male to increase the number of eggs he fertilized dramatically, compared with fruit fly males mated to only two partners, or one partner. In females, however, the second and third male mate made no difference; females who mated with one partner laid the same number of eggs as those who mated with two or three.

The result, however, was about people, not fruit flies. It was that "the male is eager for any female, without discrimination, whereas the female chooses the male." In other words, the behavior of maleness and femaleness stereotypically transcended the evolutionary gulf that divides flies from people.

Now, of course, one can't gauge "eagerness" in a fly with any great precision. But the results speak for themselves: a male fly's reproductive success is aided by his ardor, while a female's is not aided by hers.

The key question, however, is this: Have you discovered a fact about males and females, or a fact about *fly* males and females?

One way to find out, of course, would be to ask whether flies have physiological specializations that humans don't, which may facilitate the result you observed in the flies. That might tell you you're looking at the facts of specifically fly biology, not transcendent properties of maleness and femaleness. And sure enough, they do have such specializations. Fruit fly females have an organ known as a spermatheca, which functions to store sperm, so that after one mating they can fertilize as many eggs as they want. They don't need to mate again to get more sperm.

Human females lack a spermatheca. That tells us something important. It tells us that the result you obtained harmonizes well with fruit fly biology. But since human biology is different, it follows that the result should not be directly extendable to humans.

Unfortunately, that was not the lesson learned by the scientist. Nor was it the lesson appreciated by several zealous students of sociobiology, which studies the biological roots of human behavior, whether or not they exist.

OF WORMS AND MEN

Molecular genetics seems to be revolutionizing all aspects of our lives, at least to judge from the headlines. Consider the discovery that a single gene mutation in the tiny nematode worm converts it from a solitary eater to a gregarious eater, reported in the *New York Times* on September 8, 1998, with the titillating heading "Can Social Behavior of Man Be Glimpsed in a Lowly Worm?" Oh, sure, reported the *Times,* such a suggestion is "unwelcome"—the implication being that it's politically incorrect, rather than just bloody stupid and a thoroughgoing disgrace to the good name of science.

A harsh judgment?

Once again, let's start with nematode worm biology. The animal is technically known as *Caenorhabditis elegans,* or for the classically challenged, *C. elegans.* It has been a great boon to developmental biology, for its growth and development from a fertilized egg is better understood than that of any other organism. The major reason is the biological simplicity of *C. elegans.* It has, for example, 302 nerve cells.

For its behavior to be directly translatable to human behavior, those would have to be really, really good nerve cells, for that is several billion fewer than will be found in you or me.

More than that, what really is the relationship between "sociality" in worms and in humans? One can always compare animals and people. The owl is wise, the fox is clever, the ant industrious—Aesop popularized it in European literature, and cultures all over the world make similar associations in their oral lore. This is anthropomorphism, great as song and story, dubious as science. The association here between animal and person is analogical, metaphorical. You're not saying they share a common biology, but rather that they are symbolically alike.

By the middle of the nineteenth century, anatomists had recognized that there was another kind of comparison to be made among different species. It was more than a symbolic association; it seemed to be something deeper, a fundamental correspondence of structure. The basic similarity between the structure of a bird's skeleton and a human's skeleton had been noted in the 1500s, but now it was given a name: homology.

Homology is more than symbolic similarity; more than a common name for two basically different features (like the wings of a sparrow and a dragonfly). And it was Darwin who explained it: homology is a correspondence of features due to common descent. Unlike analogous features, which are superficially similar (like the color and body form of sharks and dolphins), homologous features are often superficially different but similar in deep structure (like the forelimbs of a horse and a squirrel).

Which brings us back to "sociality" in worms and people. Is this just a word game, in which case it's a waste of time if your goal is to throw light on human behavior? Or is it the same thing, a deep biological correspondence?

Unfortunately, it's not easy to tell in biology. There is no official, foolproof test to enable you to tell.

But we can make a reasonable guess. Look carefully at a ½-centimeter-long nematode worm and at your mother. How many similarities do you see? Is it reasonable to think that if "legs" aren't homologous (nematode worms don't even have them, after all), "sociality" would be?

Of course not; "sociality" in worms is like "slavery" in ants. Entomology can tell you no more about the enslavement of the Middle Passage than it can tell you about the enslavement of iron filings by a magnet. Helminthology can tell you no more about a family gathering at Thanksgiving than it can tell you about a family gathering of protons in the nucleus of an atom.

It may sound like science—it's got experiments and genes—but it's a science of metaphorical, not of biological, connection. That makes it nonscience.

And pretending that the nonscience is science is pseudoscience.

The first question we need to deal with if we want to study the genetics of variation in human behavior is: Who even *has* the behavior, the phenotype, we want to study? Suppose we are interested in the genetics of aggression. Where do we begin to look? Boxers are aggressive, but actually it's just a job to them. After the final bell, they usually hug each other. The single act that killed tens of thousands of people at Hiroshima was merely the flip of a switch. Not an especially aggressive act. Except in context.

The second question is the relationship between the range of what is normal and what is abnormal. After all, aggression is a part of the behavioral repertoire of genetically *normal* people. It is specifically the *context* of aggressive behavior that defines it as abnormal. Now, human genetics is driven by medical concerns and by biochemical techniques. As a result, we cluster virtually all of the genes we *map* into two categories, with only a little overlap. First, we map genes that have little effect on the body and that are responsible for the production of a specific molecule, characterized well *only* at the biochemical level. In this category are genes like those that make hemoglobin, those that make enzymes, those that code for the blood groups. Here, very often, variation is detectable, but the variants are generally equivalent. One can have type B blood or type O blood, after all, and still be normal.

The other category comprises those genes that have a detectable effect on the organism, the body at large. This class is composed almost entirely of diseases. We map genes for the body's breakdown: the cystic fibrosis gene, the muscular dystrophy gene. This is actually a complex shorthand. The fact is, there is no gene whose function is to *cause* cystic fibrosis. Rather, what we know is that when the functioning of a certain gene (whose function and physiological role we don't understand) is compromised, cystic fibrosis results.

The point is: *We don't map the genes for noses.*

That is very important: noses are genetic, noses are variable, and, accidents aside, noses are all normal. Because of the way we identify and name genes, we develop a map of the human genome that is very morbid: a list of diseases, a genetic map of road hazards. But we don't even have a way of conceptualizing the genes for the normal development of different bodies.

It is very easy to confuse pathology for normalcy, and vice versa. You don't necessarily learn about the normal function of an object from its breakdown. Observing the fumes that result from a broken fuel injector in your car's engine will tell you neither what the fuel injector actually does nor what exhaust fumes are normally supposed to be like. Likewise, observing pathological variation in different people doesn't necessarily tell you anything of value about normalcy.

Consider height. We know that there's genetic and environmental causes for variation in height. Tall parents tend to have tall children. On the other hand, in developed countries, people are generally bigger than their parents, whose genes they have. If you've ever looked at medieval armor, it may have struck you that bold and manly Sir Lancelot would have been physically puny by modern standards. In addition to genes, diet and lifestyle have also contributed to making people tall.

But let us focus on the genes for height. No such genes have been found or mapped, but it has been estimated that there are about eight of them. And maybe there are. The crucial thing about them is that they contribute to the range of *normal* height—the differences within the vast majority of the world that is between, say, four and a half feet and seven and a half feet tall.

There are, of course, a few people with genetic variants that affect their bones so that their body growth is abnormal, for example, the condition known as "achondroplasia," which causes very small stature. That gene is well characterized. It is a gene, and it affects height. But it is *not* one of the eight genes postulated above. Why? Because it does not contribute to the *normal* range of variation in height; the variation it produces is pathological. Achondroplastics are abnormally short. We learn nothing about "the height genes" from the study of dwarfism, because the genes governing the range of normal height are independent of the genes for rare dwarfism.

Likewise, Lesch-Nyhan syndrome is a rare condition in which a biochemical pathway (purine metabolism) is disrupted by a mutation in a gene that produces an enzyme called HPRT (or HGPRT, hypoxanthine-guanine phosphoribosyl transferase). Then physiology intervenes, and the phenotype that emerges is macabre compulsive self-mutilation; these unfortunate boys bite their fingertips and lips off and must be forcibly restrained.

But what does this tell you about nail-biters (like me), or rituals of scarification, or body piercing, or about any violent or mutilatory behavior in normal people?

Nothing.

The pathological genetic condition is entirely independent of the phenotypes of normal people. Yet to showcase the importance of behavioral genetics in *Scientific American,* a practitioner brandishes that very disease: "[A] neurogeneticist at the National Institute of Alcohol Abuse and Alcoholism . . . listed several factors he says are incontrovertibly linked to violent behavior. The gene that causes Lesch-Nyhan syndrome, which often involves self-mutilation, is one."

It's fairly obvious that the geneticist is self-interested when he tries to convince you that Lesch-Nyhan syndrome is caused by a "behavior gene." The syndrome is real, and affected children are tragic. But effectively none of the violent behavior in the human species over the span of world history—from the Battle of Vicksburg to Mike Tyson biting Evander Holyfield's ear, from the assassination of Julius Caesar to Custer's Last Stand to the My Lai massacre—has had anything whatsoever to do with Lesch-Nyhan syndrome. These are violent acts by genetically normal people and not the subject of the genetic data being invoked here.

And in spite of the news division of the journal *Science* trumpeting "Possible Aggression Gene Found," the Dutch geneticists who studied a genetic syndrome called monoamine oxidase A (MAOA) deficiency made no claim that their work had any bearing on the causes of aggressive acts in modern society, which are overwhelmingly carried out by genetically normal people. In fact, Han Brunner, the senior author of the MAOA study, made his thoughts very explicit in print: "Although genetic studies cannot explain *why* impulsive aggression occurs, they may help to improve our understanding of *how* impulsive behavior happens."

Fair enough.

Lesch-Nyhan syndrome and monoamine oxidase deficiency produce abnormal violent behavior. Violent behavior in normal people in ordinary or extraordinary contexts (sports, crime, discipline, politics, etc.) is untouched by the genetic knowledge about those conditions. But it's the "normal" violent behavior that we're interested in when we hear about "genes for behavior." Why, then, was his work reported as an "aggression gene"? Why does it get trotted out, for example by sociobiologist E. O. Wilson, in his recent book *Consilience,* as an example of the influence of "genes on behavior"—when even the geneticist who did the work doesn't interpret it that way?

What on earth is going on here?

Welcome to behavioral genetics, where the social and natural sciences square off and we learn that nature really does beat nurture. At least, if you believe behavioral geneticists. But then, of course they'd say that; they're behavioral *geneticists,* after all. And their work, however tentative or irrelevant, can then be uncritically invoked by people who hold concordant social and political beliefs about the innateness of social problems because it appears to support them scientifically. It's thus in the interests of both groups to get you thinking that behavior is genetic. Just as it is in the interests of a Buick salesman to convince you that Buicks are the best cars on the road. And maybe they're right. But on the other hand, would they tell you if they weren't?

BETWEEN NATURE AND NURTURE

Behavioral genetics is where science and politics converge. *The Bell Curve* was a 1994 best-seller that argued that since differences in IQ are innate, we should cut social programs, because they cannot advance people above their intellectual potentials. The authors of that book certainly appreciated a significant relationship between heredity and politics.

If the heredity of behavior is politically loaded, then obviously it needs to come under a special degree of scrutiny, or else work of dubious quality may be used to influence public policy and thereby degrade people's lives. And certainly if the history of anthropology has shown us anything, it is that science can easily be invoked to naturalize social inequalities, to make it seem as though they are nature's fault (and not the fault of greedy and evil people), and thus discourage any attempt to make the world better. And that's bad both for people and for science.

We began by distinguishing normal from pathological behavior. But it's not always that easy. The vast majority of people in the world, for example, are right-handed; fewer than 10% are left-handed. We don't know why. But the fact that left-handedness is fairly rare doesn't necessarily mean it is bad, or problematic, or even biologically abnormal. Culturally, however, this deviation from the majority is often very significant. There are many symbolic associations with left-handedness, which take an infrequent biological condition and construct absolute dichotomies around it. Lefties are gauche, not adroit; the left is politically unstable; the left is not the seat of honor. Lefties are sinister, not dexterous. The left is widely associated with the west, with sunset, with the dark, with cold, with death, with evil—as anthropologists have been recognizing since shortly after the turn of the century. Students in earlier generations were often forced to write with their right hands, even if their natural inclination was to use the left.

Somehow, the collective processes of the human mind have constructed a spiritual map of the opposed forces of the universe and inscribed them upon the human body. We don't know why or how this came to pass. What is clear is that an obscure and uncommon biological datum has become a subtle and powerful worldwide symbol of abnormality, deviance, badness—although there is nothing intrinsically bad about being left-handed. It is simply different from most people's inclinations.

One could search for the genes for left-handedness, and maybe some exist. But they would not explain the *meaning* of left-handedness, the significance of the biological fact. More likely than not, much left-handedness is developmental, not genetic at all, since identical twins are commonly discordant—in spite of its obvious "innateness" and the subtle differences in brain structure that accompany left-handedness. Of course, handedness can be taught, and there are degrees of it, in spite of the cultural dichotomy we impose.

Studying the genetics of left-handedness probably wouldn't tell us much about its etiology, simply from what we already know; it certainly wouldn't tell us anything about the significance of the phenomenon; nor would it tell us what to do about it, if anything. Left-handedness is either something you accept culturally as aberrant, dangerous, and requiring corrective measures; or something you live with. And after living with it for a while, you don't even notice it.

THE GENETICS OF HOMOSEXUALITY

One of the outstanding hallmarks of human evolution is the extent to which our species has divorced sexuality from reproduction. Most sexuality in other primates is directly associated with reproduction. Lemurs are only sexually active for a few weeks in January. A male chimpanzee is rarely sexually active except when stimulated by the presence of a fertile female, who displays her swollen, purple genitalia. Whether the actual cues are visual, olfactory, or behavioral—more likely, some combination of all three—is not precisely known, but it is clear that our closest relatives are not generally active sexually unless there is a fertile female involved.

The principal exception seems to be the pygmy chimpanzee or bonobo, also known as the "rare and elusive" pygmy chimpanzee or bonobo. They appear to be far more sexually active than their closest relatives, the common chimpanzees, and also appear to be sexual in nonreproductive ways—famously, in females engaging in same-sex genital stimulation. This tells us far more about chimpanzees and bonobos than about humans. In any case, the bonobo appears to be exceptional—virtually all other primates are sexually active principally as a reproductive activity.

The human difference is the extent to which we have differentiated sexuality from reproduction. When you consider the range of human sexual experiences—postmenopausal, during menstruation, during pregnancy, oral, masturbatory, et cetera—it becomes clear that a surprisingly small proportion of human sexual activity is in fact the result of the conjunction of a fertile man's penis and a fertile woman's vagina. In this light, homosexual activity seems hardly in need of explanation. It may be viewed as simply another item in the long list of nonreproductive human sexual experience. Presumably, it is inscribed in our genes to the extent that "lots of nonreproductive sex" distinguishes us biologically from our close relatives.

It may be somewhat surprising to hear that it is only in the past few centuries that "homosexual" began to be applied to people rather than to sexual acts. In ancient Greece, as is well known, a relation of apprenticeship often entailed sexual activity. In many cultures, homosexual acts take place willingly, and happily, between partners who are not "homosexuals," and who are expecting thereby neither to reproduce nor to marry. They are relevant as examples of the multiple meaning of sexual activity among humans, and as a demonstration of the cultural construction of "the homosexual."

Certainly human sexuality is far more complex than a division into two *words*—homo- and heterosexual—would suggest.

Of course, there is also a significant cultural matrix into which the genetic basis of human sexuality is cast. Germany in the 1930s strove to deal with the "problem" of homosexuality by locating it to defects in the constitutions of individuals and exterminating those with the defect. This created strong pressure, in reaction, to consider homosexuality a learned behavior, for this understanding would render the extermination program ineffectual—one could kill homosexuals, but homosexuality would continually arise. By the early 1960s, the conventional wisdom was that homosexuality was learned, which raised the question of whether homosexuals should be allowed to teach, on the presumption that they might be teaching buggery. This created pressure to see the phenomenon the other way—straight parents shouldn't worry about homosexuals teaching pupils homosexuality, for it is innate.

Certainly, many people feel as though they have "always known" they were gay, and that their lives could never have been lived any other way, for it is innate, rooted in their essence. This is a classic philosophical issue, essentialism, in which a whole person, a developmentally complex entity, is defined in terms of one innate quality ascribed to him or her.

As an issue of molecular anthropology, however, we may start out by asking whether there is an empirical scientific basis for "the homosexual," as opposed to "homosexual acts performed by ordinary people." We may immediately observe that by isolating "the homosexual" as a being fundamentally different from, and opposed to, "the heterosexual," we are setting it up specifically as a pathology to be studied, like someone with cystic fibrosis or rickets. Is "it"—is that difference—caused by genes or environment?

Three sets of data have recently been advanced as showing that homosexuality has a genetic basis. The first, a study of brain structure; the second, a study of the concordances of twins; and the third, a study of the concordance of genetic markers in siblings. Let us say for the sake of argument that they are valid. Bearing in mind that we have a strong tendency to regard visible genetic traits as diseases, because those are what we map, we create the possibility that homosexuality might be treated genetically *as if it were a disease.*

And in a recent review in the journal *Science,* that's exactly what we see. In a chart listing behaviors ostensibly mapped to the genes, the authors give: "mental retardation, Alzheimer's, violence, hyperactivity, paranoid schizophrenia," and nestled snugly in between "alcoholism, drug abuse" and "reading disability" is "sexual orientation." Regardless of one's own views on sexuality, it is not like cystic fibrosis; and it is very far from a value-neutral reading of data to list it this way—as a pathological phenotype.

"Genes Are Tied to Homosexuality and Schizophrenia," a *New York Times* callout says. Is homosexuality like schizophrenia? I don't know, but associating them in this manner is very much a culturally loaded statement.

At root is a very simple problem for behavioral genetics. Scientific analyses sometimes involve unarticulated cultural assumptions, which are imbued with the authority of science by virtue of being embedded in scientific analyses. These have to be confronted and scrutinized, for they are potentially harmful and stand thereby to taint science. If you can't do science without harming, victimizing, or stigmatizing people, you shouldn't be doing it. The burden of undertaking something controversial is to do it in such a way that you don't hurt anybody.

Unfortunately, it is invariably the people most socially marginalized who are the ones whose behaviors are different, and therefore interesting to study, and therefore easiest to stigmatize. To look at the genetic basis of homosexuality through the eyes of a molecular anthropologist, we want to know two things: (1) Is this an objective biological feature that we are studying genetically? (2) What's the genetic evidence?

We already have seen that there is a great deal of construction that goes into the feature of "homosexuality"—not the least of which is the imposition of a dichotomy upon the spectrum of human sexual behaviors and attitudes. That tends to militate against the possibility of any simple genetic models being applicable. But what about the actual data? Are they convincing?

The first line of evidence for a genetic basis of homosexuality came out on August 30, 1991, when *Science,* the most prestigious scientific journal in America, published a short paper by neurobiologist Simon LeVay entitled "A Difference in Hypothalamic Structure between Homosexual and Heterosexual Men." Comparing the brains of dead gay male AIDS victims to those of dead straight men and dead straight women, LeVay found that a tiny region of the hypothalamus (INAH 3) was larger in men than in women and larger in straight men than in gay men. Although Le Vay discussed homosexuality as "biological," he did not at that point suggest that his work indicated that it was genetic or innate. In fact, he said explicitly that "the results do not allow one to decide if the size of INAH 3 in an individual is the cause or consequence of the individual's sexual orientation." But to the *New York Times,* he made it clear that he thought it was the cause, suggesting "that the hypothalamic segment could be responsible for inspiring males to seek females."

A *Newsweek* cover story, "Is This Child Gay?" also featured LeVay's work and characterized him as a "champion for the genetic side."

But where's the genetics?

LeVay studied brains, and the key assumption promoted in the public arena is that brain structure is a direct result of genetic instructions. But it isn't. The brain grows and develops interactively with the experiences of the person; brains are not reliable surrogates for studying genes.

Unfortunately, LeVay never followed up this study, and other brain analyses have found his results to be very equivocal.

The most interesting aspect of the study is that, aside from the technology, it is conceptually very unmodern. The logic is that the brain is a surrogate for the genes, and that a consistent difference in brain structure implies an innate basis for the thoughts associated with the brain. That's entirely wrong, virtually out of a different century, and affords an excellent illustration of a simple rule of modern molecular anthropology: *Genetic conclusions require genetic data.*

The second line of evidence for the genetic basis of homosexuality was a study of twins, showing a high concordance for sexual orientation between identical twins. Twin studies have a long and rather sordid history as surrogates for genetic studies. The most notorious were those of Sir Cyril Burt, who found identical twins raised apart to be incredibly similar in IQ scores—so similar, in fact, that the statistics describing their similarities didn't change even when his sample of twins tripled. His power in the English psychology community was so great that nobody dared to challenge him, but upon his death in the 1970s, it was found that he had invented collaborators, ghostwritten book reviews under their names, and generally transgressed virtually every boundary separating the credible scientist from the wacko mad scientist.

Twins are, of course, very powerful cultural figures, the subjects of old mythologies and many hokey novels. In theory, they should be good genetic guinea pigs for nature-nurture experiments, but they aren't in practice. For example, the homosexuality twin study found

52% of identical twins concordant for homosexuality, as opposed to a control group of adopted siblings, only 11% of whom were both gay.

Of course, that leaves 48% of identical twins with different sexual orientations. And the 11% is a much higher estimate of homosexuality among unrelated people than is generally taken as a reliable estimate of homosexuality "out there."

But more significantly, this is a very crude comparison. The same study showed that fraternal twins were concordant 22% of the time, which seems to fit the idea that it is genetic (the value is less than for identical twins and more than for unrelated people); but ordinary nontwin brothers were concordant for homosexuality only 9% of the time. However, genetically, fraternal twins are simply siblings born at the same time. How could the concordance rate for fraternal twins be more than twice as high as the value for brothers generally, unless the study is actually revealing a strong effect of twinship, rather than of genetic identity?

Twins tend to be treated similarly, and tend to regard themselves as more similar to each other than ordinary siblings—identical twins even more so. This study has not managed to isolate the genetic similarity of twins and bracket it apart from the "overall" similarity of twins. The fact that they are very concordant for a particular trait consequently tells us nothing about genetics.

A second simple rule of molecular anthropology: *Similarity among relatives is not necessarily, and often isn't, genetic.* See Rule #1.

The third study purporting to show that homosexuality is genetic is the only one, oddly enough, actually to be based on genetic data. Dean Hamer, a glib and charismatic researcher, led a team that found an association between a tiny variant segment of the X chromosome (called Xq28) and male homosexuality, published with considerable fanfare in *Science* on July 16, 1993.

At least here we have some genetic data.

But what is the nature of these data and how convincing are they for establishing a "genetic basis" for homosexuality?

As already noted, modern genetics relies on establishing a cause-effect relationship between a functional bit of DNA and an observable trait. But this study didn't. It found that brothers who were gay tended to match at this genetic region.

But what is a match? Crucially, the scientists isolated no gene there, and no physiological product affecting sexual orientation is known to be made there. The claim made by Hamer and his colleagues is simply that they found this region to be grossly more similar (83% matching of small genetic marker regions) in a specific sample of gay brothers than at random (50% matching). The result hinged not on a mechanistic analysis but on the statistical difference between the frequency of gay brothers with similar genetic markers in a specific chromosomal region and the random expectation.

The cultural meaning they imparted to the result they reported was that "chromosomal region" implied "functioning gene," which in turn implied "control of trait."

Assuming, of course, that the association was real to begin with. The tricky design of the study makes it very sensitive to a few families matching or not, because the key scientific question is *not* "Do we have a gene for homosexuality?" but a surrogate question: "Is our 83% result sufficiently different from 50% to be meaningful?" A follow-up study by a different group found no such difference at all. Another follow-up by the original researchers found the difference now to be 67%, rather than 83%, quite a bit closer to the 50% expected at random.

Nevertheless, best-sellers were written; careers and fortunes were made. "Born Gay?" *Time* asked on July 26, 1993. And the "gay gene"—which has never subsequently been found—entered the popular mind as a fact of science.

For me, however, the most interesting aspect of the study was the scope of the actual claim. How much homosexuality did these researchers believe they had actually explained with their study? From the publicity, you might expect the figure to be 90%. Or perhaps a more conservative 70%—perhaps they had explained over two-thirds of the homosexuality in our species, which would certainly merit headlines.

In fact, however, when I posed that very question at a conference in 1996, the answer was very different. It came in two parts. First, the result, according to the researchers, was ostensibly only about *male* homosexuality and had no relevance at all for female homosexuality; and, second, they believed they had explained about 5% of male homosexuality.

Five percent.

If we make a simplifying assumption that male and female homosexuality exist in the universe in equal proportions, then *at best*— assuming that homosexuality is a property of a person, not of an act, and assuming all the statistical issues raised are invalid, and assuming there is actually a gene there—they would have accounted for 2.5% of homosexuality in our species.

The third rule of molecular anthropology: *There is no science other than behavioral genetics in which you can leave 97.5% of a phenomenon unexplained and get headlines.*

That is the most obvious indicator of the cultural power and meaning of this work, and why it needs to be considered very carefully and regarded very skeptically. Virtually any claim, no matter how ridiculously small, can grab headlines. The question is not, "Do you believe homosexuality is genetic?" After all, the Constitution of the United States guarantees you the right to believe anything you want. The question is, "What have we actually shown scientifically about it?"

And the answer is, almost nothing.

THE SCOPE OF BEHAVIORAL VARIATION IN HUMANS

The fact is, we already know a great deal about behavioral variation in the human species. We know that virtually all the detectable behavioral variation *between* groups of people is the result of cultural history. Why? Studies of immigrants, acculturation. The fact that most of your ancestors three, four, or five generations ago spoke different languages, ate different foods, had different aspirations, led lives entirely different from yours today.

Odd as this may sound, Americans today are very behaviorally homogeneous. In the kinds of foods we regard as edible, the sounds and gestures we regard as meaningful, in terms of making some sense of our lives, what we wear, the composition of our diet—the entire fabric of our daily lives, no matter how diverse it seems, actually encompasses a very narrow range of the possibilities of variation realized elsewhere in our species.

The point, then, is that the bulk of human *behavioral* variation is *between* groups and is nongenetic. We know this. Imagine trying to communicate with your own ancestors five generations ago, much less with the ancestors of your neighbors. Tremendous amounts of behavioral change can occur within a few generations, either in the same place (due to technology and social forces) or by the process of immigration and internalizing a different set of normative ideas and values. And all in the absence of genetic change: cultural variation is the bulk of behavioral difference in our species, it comprises the differences between groups of people, and it is, as far as we know, entirely nongenetic.

Within groups, there may well be genetic variation. Maybe there were genetic differences at work constructing the brains of, say, Arsenio Hall and Al Gore. But the behavioral differences between them are dwarfed by those between *either* of them and a Sherpa from Nepal. Both American men wear pants, shirts, and underwear, eat with knives and forks, have similar ideas about what is edible, what acceptable standards of behavior are for them in various roles as citizens, sons, fathers, husbands, professionals; they speak the same language, find meaning in similar classes of gestures and sounds, and find similar things funny, offensive, or revolting. None of these similarities can be taken for granted when comparing people from different places or times. In the spectrum of human behavior, the great bulk of its differences are to be found in contrasting the myriad lifeways of the peoples of the world (which are rapidly being reduced, of course), which are the result, as far as we can tell, of differences of nothing but history.

But this raises the question, exactly what are we looking for when we talk about behavioral genetics? *We are looking for a hypothetical genetic component to account for a very narrow bandwidth of behavioral diversity in the human species.*

Is it there? That depends on what your standards of evidence are. So let us think about modern science.

We now know that science is more than just the collection of facts. It's a process by which some sort of factualness is established in a specific atmosphere of cultural views, class biases, time and place, and conflicting interests. And human genetics is one area where that is most obvious.

After all, when a genticist says, "Social problems are genetic in origin," that might be a true statement about the universe. But it definitely is a grant proposal, a request for funding. If it happens to be true, it's good for business; but merely saying it's true may be just as good.

Now, scientists are citizens and are as much a product of their time and class and culture as everyone else. Consequently, they are encumbered with similar ideas to everyone else about their place in the world, about women, blacks, Jews, sexuality, the poor, and so on.

However, saying, "Nine out of ten doctors smoke Lucky Strikes" carries more weight than "Nine out of ten grocery store checkout clerks smoke Lucky Strikes." Madison Avenue knew that generations ago. Scientists are smart; they are scientific. Scientific statements are *authoritative.* That places a burden on scientists to be careful in their social pronouncements and raises a question about the implications of their being wrong.

RESPONSIBILITY AND THE PRONOUNCEMENTS OF HUMAN GENETICS

Madison Grant was a well-heeled, Yale educated New York lawyer with a broad moustache and an obsession with biology. Along with his friend Theodore Roosevelt, he helped to found the New York Zoological Society. And his 1916 book *The Passing of the Great Race* is a classic of American popular social thought.

It argued that America was imperiled genetically by the short, swarthy poor people from southern and eastern Europe immigrating in large numbers; and that the way to save America was to enact legislation to sterilize them and restrict immigration—specifically on the basis of the American gene pool. Grant proposed the elimination of the weak or unfit, "beginning always with the criminal, the diseased and the insane, and extending gradually to . . . worthless race types."

If this sounds un-American, or unmodern, of course it does. That's what makes it interesting. Madison Grant wasn't a Nazi, because in 1916 there were no Nazis. But by the mid-1920s, federal immigration restriction laws were indeed on the books in the United States, and state sterilization laws were upheld by the Supreme Court's 1927 ruling in *Buck v. Bell.* These were popular ideas in 1920s America.

I would like to be able to report that scientists rose with indignation to bash this book. A few, like anthropologist Franz Boas, did. But most didn't. For example, the book was reviewed by a geneticist from MIT in *Science,* and he glowed about it. The book was good for business if you were a geneticist, because it was fundamentally about the importance of good heredity in American social life. And the fact is, it didn't misrepresent contemporary genetics much, if at all. It was an application of modern ideas about genetics toward the amelioration of social problems; which is why it was reviewed favorably, and why Grant himself received fan mail from politicians as diverse as Theodore Roosevelt and Adolf Hitler.

What Grant wrote, a manifesto of the "eugenics" movement in America, is actually what most geneticists thought. It's basically what they taught and what they wrote themselves. Thus, they were able to promote it as the scientific view of social problems. If you opposed it, you were branded as being against science, against progress, against modernity.

It would be a mistake to underestimate these scholars, for there was some very seductive scientific reasoning here. Madison Grant's scientific inspiration was the work of his close friend Charles Davenport, one of the two most prominent geneticists in America at the time, who argued about social problems from a scientific, materialist genetic perspective. The genes or "germ-plasm," argued Davenport in his 1911 book *Heredity in Relation to Eugenics,* code for the brain. The brain is the seat of the mind. The mind contains the thoughts that lead to actions.

Bad thoughts and actions, therefore, are caused by bad genes.

In other words, immorality can in principle be controlled biologically by regulating the births and deaths of immoral people.

It's *not* stupid or illogical, but it is dead wrong.

In fact, you can pick up literally any textbook of genetics from the 1920s and find those ideas. Here's a particularly interesting excerpt from the first edition of a very widely used textbook from 1925: "[E]ven under the most favorable surroundings there would still be a great many individuals who are always on the border line of self-supporting existence and whose contribution to society is so small that the elimination of their stock would be beneficial." That is another distinctly un-American thought—eliminating people's stock on the basis of their limited contribution to society! Of course, it's just the poor, but remember Madison Grant—we *start* with the poor, and work our way out to worthless race types. The thought should be unsettlingly reminiscent of authoritative scientific pronouncements we associate with the Germany of several years later.

The problem is perennial and simple: When science justifies violating civil and human rights, it doesn't much matter *which* people or *which* rights. And it's always going to be cheaper to kill people than to operate on them, as we also know in tragic retrospect.

The history of that textbook is instructive: the entire chapter was deleted from the second edition in 1932. Now, of course, between 1925 and 1932, the market crashed and *most* people's stocks were eliminated, which demonstrated pretty clearly that wealth was not necessarily a good predictor of genetic value, something geneticists had hardly realized before that.

The point of this history lesson is that the facts did not speak for themselves.

Scientific answers to social issues implied political action. Scientists such as Davenport knew this and strove for it. It's easy to see their wrongness in hindsight. But that doesn't help the Americans who were sterilized against their will or those who couldn't escape the Nazi regime because of the newly restrictive U.S. immigration laws.

Do today's geneticists bear a burden of responsibility for their predecessors' wrongness? If so, what is it?

I think the answer is yes, and here is what I think it is: the responsibility to understand what the errors were, to ensure that they don't happen again. In science, after all, you don't have the right to make the same mistakes over and over—you only have the right to make new and creative mistakes.

On the one hand, of course, nobody is seriously talking nowadays about breeding a better form of citizen. But, on the other hand, we also know that crime and poverty are not biomedical problems. Crimes are defined by social convention, and poverty is the result of social, not biological, forces. The act of taking a human life may be justified if it is done in time of war, or in self-defense, or to appease the gods, or in some other context. It is the context, not the mere action, that defines it as a crime. The crime is specifically taking a human life *in the wrong context*. Thus, genetics is an inappropriate arena for contemporary efforts to deal with crime.

That is what brings us back to behavioral genetics.

When the National Institutes of Health decided to sponsor a symposium exploring the genetic basis of crime in 1993, it encountered an extraordinary backlash. Some trivialized this backlash as "political correctness," suggesting that it was mindless anti-modern, anti-science emotionalism, but others appreciated that critics of the eugenics movement—unfortunately largely unheeded until it was too late—had met with a similar reaction. The fact is, whether crime has a genetic basis is not a soluble genetic problem, but the idea that it does have such a basis can easily give rise to extraordinary claims, which may enter the public consciousness and even shape public policy.

So we had better be damn sure about them.

A BROADER PERSPECTIVE

A perspective from molecular anthropology would dictate that while there are empirical data to be collected, we are nevertheless not starting at ground zero. There is an abundance of data—both scientific and humanistic—on the problem of the genetics of crime. Since crime isn't an objective entity (theft is contingent upon notions of property rights; murder is contingent on notions of human rights), it would be surprising to identify it in the genome!

We know that scholars over several generations believed that they had found an organic basis for crime, only to have fundamental flaws revealed in their research. Before the days of DNA analysis, Harvard's Earnest Hooton measured the skulls, faces, and bodies of thousands of prison inmates and compared them to model citizens to see whether criminality was identifiable in the constitution of the criminal, marked in his body. And Hooton concluded that the criminals were indeed slightly different physically on average from volunteer firemen.

But it turned out that there were small differences in the composition of his criminal and control populations. For example, the criminals averaged four years younger and twelve pounds lighter than the firemen. These differences were, of course, minor, but four years can make a difference in one's face and body. So the fact that he could find subtle physical differences between his criminal and control populations did not mean that they differed because of criminality, but perhaps simply that they were not perfectly matched.

Hooton's point was unconvincing and was thus valueless as scientific evidence. Science, after all, is supposed to be *convincing* knowledge. Harvard University Press published his findings as volume 1 of *The American Criminal,* but never released a second volume. Although it was useless for the scholarly community, Hooton packaged the work for a popular audience under the title *Crime and the Man.*

The 1960s saw another vogue in criminal-genetic associations. The Y chromosome has but one major trait associated with it—maleness. Maleness is genetic, but is very difficult to distinguish from masculinity, the cultural construction of maleness. So the Y chromosome is the focus of both genetic and cultural meanings, which may have a considerable impact on the life of the 1 in 1000 boys born with an extra one.

The geneticist Patricia Jacobs and her colleagues found in a 1965 study that 3.5% of the inmates in a mental/penal institution in Scotland had an extra Y chromosome—far more than the percentage of XYYs "at large." Although they carefully stated that they did not know whether aggressive behavior was the cause of XYYs' over-representation in the institution, they nevertheless made it clear that they had undertaken the investigation on that assumption.

The logic, however, is cultural: the Y chromosome determines maleness; men are more aggressive than women; so men with an extra Y chromosome should be extra-aggressive. Nevertheless, by that reasoning, the XYY men should also spend more time watching football on television, burp more frequently in public, and be especially good at math. But those traits weren't examined. The men were simply found to be disproportionately institutionalized.

The culturally loaded interpretation of such findings began to have its effect. In 1968, a false newspaper story gave an extra Y chromosome to the mass murderer Richard Speck. In 1971, Bentley Glass, a distinguished geneticist, called for a new and presumably benign eugenic program to help "rid us" of "sex deviants such as the XYY type." And the third, 1992 installment of the *Alien* movie series takes place on a remote penal colony in outer space, reserved for the most dangerous criminals in the galaxy—the XYYs.

But more detailed studies of XYYs were showing that they were not hyperaggressive and supercriminalistic at all. The over-representation in mental-penal institutions was real enough, but XXYs were also overrepresented, even though they had the opposite of the extra "violence chromosome"! The people with abnormal chromosome configurations were apparently more prone to being caught than to committing crimes.

A screening program was soon set up to detect newborns with XYY syndrome and monitor them, but the project's design confronted researchers with a catch-22: unless they deceived the parents of an XYY son, they would be underwriting a self-fulfilling prophecy, because the parents would be raising the boy in the expectation that he might have criminal tendencies. Under protest, the screening program was therefore aborted. But some years later, the geneticist who had published the original study castigated "environmentalists who clearly felt threatened by the suggestion that there might be a genetic component to behavior"—as if they were the ones who had introduced social politics into science.

The problem was that the original scientists were insufficiently attuned to the social/political consequences of what they were doing. On the scientific side, the male attributes provided by the additional Y chromosome appear to be limited to height and acne. The greater likelihood of incarceration for aggressive behavior may, moreover, be a simple consequence of something well known in genetics, that adding extra chromosomal material generally reduces intelligence and survival. The Y chromosome, being very small and having few genes, permits a higher survival rate and compromises intelligence less than other extra chromosomes do. Thus, most XYYs tend to be ordinary people, unaware of their genotype and with no reason to suspect that it might be anything other than normal.

Much later, oddly enough, both Hooton and the XYY work were cited favorably in an overtly political 1985 book advancing the genetic basis of crime by Richard Herrnstein, a decade before his book *The Bell Curve.*

The problem seems to be that *saying* crime is genetic is newsworthy, when *proving* it is what ought to be newsworthy. In human behavioral genetics we find an extraordinary pattern in which claims for genetic associations to behaviors are made, along with a self-serving plea for more funding and studies to corroborate this "preliminary" result, and then a less widely reported failure to corroborate it. The examples are legion: alcoholism, depression, "novelty-seeking," schizophrenia, homosexuality. . . . When Dean Hamer's group reported finding a genetic link to homosexuality in 1993, the news made page 1 of the *New York Times;* but when another group a few years later looked for that same genetic link using the same methods, and failed to find it, they only made page 17.

The question for behavioral genetics is: What do we know now that makes this modern science? What have we gained from Davenport, or Hooton's study of criminals, or the XYY syndrome, that makes us believe that there is a biological basis of crime, and that we can study it scientifically? What do we know that they didn't? Can we show that we are not just committing the same intellectual mistakes over and over again—only now aided by different and newer-fangled technologies?

The tragedy is that usually we can't. Between the self-interested pronouncements of behavioral genetics and the conservative social and political interests who recognize that an innate basis for behavioral deviance complements their agendas, we really have little to guide us in locating the ostensible scientific basis for understanding human behavior.

One often hears the challenge "to find something wrong with the study"—to shoot it down. You have criticized it, you've raised questions, so the challenge goes, but you haven't refuted it. The answer is simple: *I don't have to.*

The burden of proof in science always falls on the claimant, not on the critic. The challenge "prove me wrong" is the classic signature of the quack and the charlatan—the one who wants you to believe that Martians built the pyramids, or that they are communicating with the dead, or that their body encases the reincarnated spirit of the Queen of Atlantis. Scientific credence requires high standards of evidence. Speculations are cheap, and you're entitled to believe whatever the hell you want, and even to make a buck off it, but if you lay claim to the authority of science for your beliefs, you must expect to be subjected to extraordinary demands.

And quit whining about it.

Sometimes we can't even identify the fallacy. Recall the Manoilov blood test of the 1920s, which could distinguish the sex, race, and sexual orientation of a person. You simply added a few chemicals, shook the sample, waited, and watched. In itself, not that incredible, although it was before the discovery of the human sex chromosomes or any way to examine them reliably. Perhaps Manoilov had unknowingly developed a crude assay for hormone concentrations?

The interesting thing is that it engendered much interest and little criticism in the scientific literature.

Hooton, the leading American student of race, would have welcomed such a test, but couldn't accept Manoilov's. It was not so much that he knew Manoilov's results to be wrong; he knew them to be *impossible.* They were so weird from the standpoint of critical anthropological thought that "right" and "wrong" didn't even seem to apply.

Manoilov's blood test seemed to make a lot of cultural sense, however. After all, blood is a powerful metaphor for heredity. We quite commonly talk about traits being "in the blood." But being a Russian or a Jew (tellingly, Manoilov assumed that you couldn't be both!) was a fact of social history, not a literal question of blood. Being a Latvian or a Pole was political history, not natural history. As Hooton summarized it in his 1931 text *Up from the Ape:* "The results of the Manoiloff test do not inspire confidence. . . . It is inconceivable that all nationalities, which are principally linguistic and political groups, should be racially and physiologically distinct."

Hooton, it should be noted, never identified a technical, methodological flaw in the blood test. To this day we don't know what that curious test was actually testing. Most likely it was simply verifying the investigator's expectations.

Another anthropological generalization about science: it's very easy to come up with results you already believe.

Manoilov's work was still being cited in genetics textbooks into the 1940s, however. And why not? It was technological, it was statistical, it was quantitative, and it was methodologically explicit. But it was nonsense.

It's a fool's errand to try and identify a methodological problem with every study making a claim about biological differences among people. What we can do, however, is integrate scientific and humanistic knowledge about our species to identify the major cultural fallacies associated with these studies, and help point us—by counterexample—in more fruitful scientific directions.

Six

FOLK HEREDITY

THE HISTORY OF GENETICS would be no different if Gregor Mendel had never been born. The field was not at all influenced by his ideas, and it wasn't until thirty-five years after his work, in 1900, that other scientists independently hit on the same results, only to find that he had scooped them by decades. In creating a new science, actually given the name "genetics" by the English biologist William Bateson in 1906, they adopted Mendel as a figurehead.

Mendel's "Two Laws," now universally memorized by college biology students, were not even formally codified as such until the American geneticist Thomas Hunt Morgan did so in 1916. Prior to that, geneticists simply wrote about "Mendel's Law" and proceeded to describe it vaguely, without the trappings of a lawlike proposition.

Today, we follow Morgan and pay homage to Mendel by recognizing two explicit scientific laws of heredity, which we usually express in abstract alphabet-soup form. The first governs the transmission of a single gene. This Law of Segregation tells us that genes come in pairs and can mask one another's effect, but only one of each pair is transmitted to offspring. The "masking" of one gene sequence by another means that a phenotype (what you appear to be) is not a reliable guide to a genotype (your genes). The same phenotype can be the result of different genotypes.

The second law, the Law of Independent Assortment, governs the transmission of different genes—or at least, of genes on different chromosomes. It tells us that because of the statistical behavior of chromosomes during cell division in the reproductive organs, and the random association of egg and sperm, Like doesn't necessarily beget Like. Organisms can have offspring with visible features (phenotypes) unlike their own. Two people with brown eyes and without cystic fibrosis can have children with blue eyes and cystic fibrosis, or either of the two traits—although they're not both equally likely.

Such formalizations constitute the core of the modern science of genetics. Subspecialties have their own laws, for example the "Central Dogma of Molecular Genetics": to wit, that in higher organisms, DNA encodes RNA, which contains the code for proteins.

Genetics is recognizable. You know it when you see it, because of the terms it uses: chromosome, gene, allele, heterozygous, phenotype. You know it because of the ways it validates itself: blots, bands, gels, sequences.

Genetics is the scientific study of heredity. While that may sound redundant, it isn't. There are other, nonscientific approaches to heredity. There is the old European idea, for example, that children inherit the attributes their parents gained during the course of their lives. While this is true of trust funds, it isn't true of physical, bodily features. After all, throughout the course of the parent's life, the features they will pass on are already sequestered away in the cells of their reproductive organs. In premodern Europe, it was widely held that an adulterous woman would find her lover's traits in children she bore many years later. The Tsonga of southern Africa hold that the mother contributes nothing to the child she bears; the Trobriand Islanders of Melanesia hold the opposite, that the mother contributes all.

Often these ideas about heredity serve to justify other beliefs, such as whether a child belongs to its mother's or father's lineage; or to whom the child is an heir. This is because there is no obvious reason to distinguish formally between heredity (as in biological transmission) and inheritance (as in social transmission).

A scientific theory of heredity, of course, finds it necessary to make precisely that distinction, and to make certain that the transmission of sickle-cell anemia is differentiated conceptually from the transmission of Aunt Minnie's silverware—although both may run in the family. Failure to do so might suggest to someone born with one of Aunt Minnie's silver spoons in their mouth that their advantages in life were simply natural, a part of their biological endowment. Or conversely, that someone born without such advantages in life deserves no more.

This was the core of the nineteenth-century philosophy called "social Darwinism." By confusing the biological fact that people aren't *identical* with the social observation that people aren't *equal,* social Darwinism developed a conservative political philosophy that justified existing social hierarchies as natural facts. The wealthy and powerful were in their rightful place by virtue of the "survival of the fittest," a phrase Darwin had not coined but had ultimately come to accept as synonymous with his own "natural selection." The political implication was that the existing social hierarchy was natural, and consequently that any attempt to alter it would represent a subversion of nature. The people on top were there for a reason—they were simply better, and deserved their position; the cream had risen to the top. Thus, ideas like welfare, labor laws, and other social programs geared toward promoting social mobility were bad and should be abandoned, because ultimately they would only fly in the face of the way things had evolved to be.

The power of this view, both as a genetic-sounding theory and as scientific justification for a political movement, is evident in the popularity of 1994's *The Bell Curve,* which made headlines, magazine covers, and talk shows because of its thesis that (as evidenced by pencil-and-paper tests) the lower classes were innately intellectually inferior to the upper classes, and that consequently nothing can be done about it. And therefore nothing *should* be done about it.

New factoids and new ethnic groups are plugged into the holes, but the theory is over a century old. And its fallacy is as simple today as it was in 1900: it may sound like a theory of genetics, but it isn't one.

It isn't a theory of the *science* of heredity. It's a theory of a different order of ideas about heredity. Cultural ideas about heredity. *Folk heredity*.

These ideas are tenacious. They sound as though they might be true. They often make sense. They sound as though they might even be scientific, because they can be dressed up in the vocabulary of genetics. More than that, like their predecessors, they are generally unthreatening to the dominant social classes, because they serve to reinforce and justify that dominance. Sad to say, scientists (even geneticists) frequently believe these folk ideas about heredity, because scientists are members of our culture and grow up internalizing its values and folk ideologies. That is why the perspective of anthropology is particularly valuable in coming to grips with the science of heredity.

The combination of anthropological knowledge and genetic data allows us to identify four cultural ideas, which translate into scientific fallacies, about human variation.

TAXONOMISM

We saw earlier how the idea that there is a small number of basic kinds of people, equivalent to subspecies, accounts poorly for the patterns of biological variation we actually find in our species and is actually a construction of social history. Humans differ from one another, but the difference is patterned locally, not continentally.

Nevertheless, it is human nature to construct difference and impose meaning on it (that is largely what we mean by "culture"), and since the eighteenth century it has become the dominant cultural view that the peoples of Africa are categorically equivalent to one another, the peoples of Asia are categorically equivalent to one another, the peoples of Europe are categorically equivalent to one another, and each is categorically distinct from the other two. This isn't false simply because of the interbreeding that has occurred between members of the different races, or simply because of the large migrations that have taken place over the millennia. It's false because of the assignment of meaning: it makes differences among Africans meaningless and differences between any African and any European meaningful. More than that, it makes anyone with *any* known African ancestry meaningfully different from anyone with *no* known African ancestry.

And those are not biological patterns.

So we can call the anthropological fallacy of treating human biological variation as if it were actually partitioned into natural subspecies "taxonomism." Taxonomism imposes qualitative distinctions between people where none exist in nature.

Geneticists have been particularly susceptible to taxonomism because the presumptive differences among groups being highlighted are supposed to be genetic. Given (1) the knowledge that races must be found in the human species and (2) the idea that they are supposed to be genetically bounded entities, it is not surprising to find that geneticists have maintained some authority in race classification for much of this century.

Perhaps the most illustrative example can be taken as the swan song of this line of argument, a 1963 review article in the journal *Science* by the distinguished biochemical geneticist William C. Boyd. Boyd begins by telling the reader, "Racial differentiation is the end result of natural selection . . . in a population sufficiently isolated genetically." But he crucially fails to distinguish here between populations (which are local) and races (which are presumably larger). If a population is all that a race is, then there are many, many human races, and the word loses its meaning. A race is supposed to be a megapopulation, of which there are few.

In fact, Boyd came to the conclusion that there were thirteen of them. But an examination of his races shows them to be repositories of cultural knowledge rather than objective tallyings of biological patterns. For example, Boyd names five races from Europe, but only one from Africa. Thus, he formally differentiates the Basques of the Pyrenees from "Mediterraneans," and "Northwest Europeans." But the tall, thin Nilotics of East Africa, the pygmies of Central Africa, and the small-jawed and flat-faced Khoisan of southern Africa merit no such distinction.

Apparently, they all looked alike to him.

And establishing the Basques as a unit of the human species equivalent to "Africans"—as if they had green skin and square heads!—is a highly arbitrary judgment. It was based on their high proportions of Rh-negative blood, which is very likely a genetical trivium here being elevated to transcendent proportions.

The point is that no formal division is objectively discernible, and, as noted earlier, we have no explanation for what it would represent in evolutionary terms if it did exist.

The problem with taxonomism is not so much that it creates opportunities for prejudice, as any classification invariably may (although that problem lies with the prejudiced person, not necessarily with the classifier). Rather, the problem is that it creates an opportunity for some very badly designed research. Thus, under the regime of taxonomism, if I wish to study some aspect of the biological diversity of the human species, I had better not confuse the "Mediterranean" people of Italy with "Northwest Europeans"—but essentially any Africans can represent that continent.

If this sounds like a silly worry, consider that just such a study of human variation was published in the prestigious *Proceedings of the National Academy of Sciences* in 1991 and generalized about the gene pool of "Africans" from a sample of Mbuti and Biaka pygmies! Africans are Africans, under this fallacy, and any is as good as any other.

But the Basques! Now *they're* supposed to be different! In point of fact, of course, a Basque in a roomful of French and Spanish people would be indistinguishable from them either genetically (Rh-negative blood can be found in people of all origins) or any other way than culturally. Basques pride themselves on their cultural distinctiveness and have a history of political separatism. They emphasize their differences from the peoples who surround them.

And that's the problem—the conflation of cultural and natural patterns of variation.

Many genetic studies have uncritically sought to analyze the "genetic distances" between "the three races," where genetic distance is an abstract measure of differences between populations, and the three races are black, white, and yellow, or Negroid, Caucasoid, and Mongoloid, or African, European, and Asian, depending upon your lexical fancy.

Two papers published by distinguished geneticists in 1974 show the taxonomic fallacy at is rawest. Stanford geneticist Luca Cavalli-Sforza showed in the *Scientific American* that genetically Europeans and Africans were most closely related and had diverged from Asians about 35,000–40,000 years ago. But using the same class of data and similar taxonomical assumptions, but different statistical tests, Masatoshi Nei of the University of Texas showed in the *American Journal of Human Genetics* that genetically Caucasoids and Mongoloids were most closely related and had diverged from Negroids about 115,000–120,000 years ago.

Obviously, the two sets of conclusions are incompatible. You don't have to be Aristotle to see that. Both scholars were and are competent, indeed brilliant, population geneticists. But actually, more likely than not, *both* were wrong.

What they lacked was a grasp of the underlying anthropology. They assumed that the familiar racial groupings had real, biological significance and used those groupings to structure their scientific questions. As the saying goes, "Garbage in, garbage out."

In fact, as noted earlier, the three categories are not even equivalent to one another. Africans comprise a diverse paraphyletic group, subsuming the ancestral gene pools of Europeans and Asians. This makes such a three-way comparison scientifically meaningless.

But even more significantly, Cavalli-Sforza and Nei were working within a framework that assumed divergence as the principal microevolutionary process. In other words, they assumed that programming a computer to *show* a branching tree representing differences among races would necessarily imply a literal branching divergence of those races. But in fact the history of the world has been far more complex, for the history of human populations is reticulated, like the capillaries of the circulatory system, diverging and fusing. The computer program wasn't representing the process accurately and it was too easy to misread the output. One of the papers even used black Americans as stand-ins for Africans—both, after all, are "Negroids."

In all, the cultural knowledge they brought to the scientific analysis screwed it up. Race is a concept derived from cultural knowledge, not from biological analysis. The problem is that our cultural knowledge seems so natural to us that it's hard to recognize it for the artifice it is.

Thus, a correspondent of mine writes that he has solved the problem of race. It isn't a social construction, he says—it is real, like an extended family. *A race is just a very extended family,* he says.

Now, as it happens, aside from race, another thing that anthropologists have studied quite a bit over the past century is the family. One of the most distinguished recent students of that institution was the late University of Chicago social anthropologist David Schneider. And one of Schneider's most interesting observations was that Americans have some strange presumptions about the family. For example, Americans take great pains to distinguish natural facts (what's objectively "out there") from phenomena that are confabulations of the human mind. And they believe that the family is based on the former, rather than the latter.

But that doesn't stand up well to critical reflection. My correspondent imagines himself at the center of a sort of familial spiderweb, with his relatives radiating outward and the boundaries of his race simply far away.

But that is just imagination, for families aren't built that way. They involve not only genetic ties but symbolic ties—paramount among the latter, of course, being the symbolic tie of marriage.

Genetic ties, in fact, form a relatively small part of what composes a family: all those aunts and uncles—some blood relatives, some related by marriage, all those step-, foster, and in-law relations; all those ex-spouses and ex-families; all those unacknowledged offspring. That is as true for the tribe of Americans as it is for any other tribe, as anthropologists have long acknowledged.

What my correspondent didn't realize is that a *family* is a social construction. The dichotomy between natural facts and cultural facts is actually a false one, for both sets of facts work together, and they are often inseparable in imparting meaning to our world. In families, there is certainly a relationship by blood, and there is considerable significance attached to it. But it neither encompasses nor is the key and overriding component of the family. After all, I am blood kin to my daughter, but not to my wife. And she is at the very heart of our family.

We create families by conventions (marriage; adopting a father's surname) that embellish the natural genetic linkages with a host of cultural associations, which are in turn no less familial for being cultural. Sometimes the genetic linkages are embellished; sometimes they are obscured.

So the spiderweb metaphor fails to encapsulate the reality of the family. It can't tell me the answers to questions that would be answerable if families were natural, genetically based, "objective" entities. Where does one family stop and another begin? How many families are there, how can I tell them apart, and how do I know which one—or how many—I belong to?

The answers aren't found in biology or genetics. They aren't "out there" as natural facts simply to be discovered. They are there to be considered and interpreted, as examples of the myriad ways in which the world's societies track and make sense of social relations by creating distinctions and imposing meaning on them.

And it is the same with race. Races aren't there as natural facts, they are there as *cultural* facts, which overwhelm and redefine the relatively minor biological component they have.

My correspondent is right. The race is like an extended family, *although for the opposite reason that he imagined!*

Taxonomism, then, is the fallacy of taking races to be units of nature. Races are real, as families are real. But they are units of cultural meaning, not units of biology. There is biology there—I am genetically related to my parents and to my daughter, and the Swiss are closer to the Italians than to the Rwandans—but they are not fundamentally natural entities like *Drosophila persimilis,* or Cetacea, or lithium, or Betelgeuse.

As noted earlier, we can, of course, make comparisons between groups of people and study their differences. The problem is invariably what meaning to assign to those differences. If we know that there are gradients, not boundaries; that human variation is patterned locally, not transcontinentally; that the extremes are not the purest representatives of anything, but simply the most divergent; that populations are invariably mixed with their neighbors, and in the last half-millennium with people from far away; and that clustering populations into larger units is a cultural act that values some differences as important and submerges others—then race evaporates as a natural unit.

The term has no technical meaning, only a colloquial one. I use the word because everyone knows what it means (sort of); but I'm not talking about anything formal when I do. I use "race" the way I use "angels" or "psychic energy."

Human differences are just that—differences. The differences between a black person and a white person may be due to many things, but if we begin by assigning them to races, then those differences become racial differences. Likewise, the difference between a group of Swedes and a group of Kenyans is only racial if you begin by imagining them to be representatives of larger, mutually exclusive, entities.

I was sitting in my office recently when a young female student came in. She wasn't in any of my classes but knew of my interests and that I had written widely on the subjects of genetics and race. Her question was simple: Could I recommend a genetic test that would tell her what she was?

How could she not know what race she was? She had fair skin, hazel eyes, and straight, dark hair. Not exactly a racial poser for the ages. Obviously, there was something here that the student wasn't communicating to me. She said her ancestry was "mixed," and I explained that so is everybody else's.

But she was in emotional pain, and she had a sad story about adoption and about not knowing her parentage. She wanted me to suggest a genetic manner of telling what her ancestry was. I explained that no such test exists. There are some genetic features that are more common in some groups than others, but they're just not racial markers—you don't have to be African to have the sickle-cell allele. Perhaps if we used several different markers, and got lucky, we could make a guess that she had some West African or Native American ancestry. But she could be black or Amerindian and lack those features, or as white as the driven snow and have them. That's just the way heredity works.

But it was clear just from looking at her that if she had any recent nonwhite ancestry at all, it wasn't much and was unlikely to be detectable genetically.

Finally, I asked the question that had been puzzling me. Exactly what did she think she was, and what did she think she was mixed with?

She obviously was greatly distressed by the fact that she did not feel comfortable without a racial identity. And "white" was certainly not adequate to her. I told her that she looked like anybody else from northern or central Europe, and that that wasn't esoteric anthropological knowledge, just common sense. And of course, she could have had a distant ancestor or two from pretty much anywhere, from a Baffinland Eskimo to a Kenyan cattle herder, but in appearance she seemed at least mostly of European descent. What was the problem?

No, there was more, she confided. She believed she might have some "Middle Eastern" ancestry.

If there were ever a time that the peoples of the Middle East were not in genetic contact with those of Europe, I explained, we don't know of it. From the start of recorded history, we hear of dispersals of peoples and migrations. They have probably been going on since Europe and the Middle East were inhabited. And certainly for the past few thousand years, the entire Mediterranean region has been in quite extensive genetic contact.

A genetic test couldn't do it. Only a *magic* test could enable you to distinguish "European" from "Middle Eastern" ancestry. She had believed so strongly in the symbolic, folk heredity of racial differences—of labeling people taxonomically—that she was grasping at the straw of genetic testing to try and help her achieve an identity she felt she lacked. An identity she felt she needed.

She left disappointed.

Aside from the sadomasochistic lunatic fringes, you won't find many people in present-day America who will stand up and cheer for racism. We grow up learning that racism, whatever it is, is bad.

Unfortunately, it is a term that is so overused that it is in danger of losing its power. The problem is that we learn that racism is bad without learning exactly what racism is. Sometimes we hear, for example, that it is racist to say that crime is genetic. If that's true, it's true only very indirectly, only insofar as crime rates correlate in America with race. It's dumb to say that crime is genetic; it's the fallacy of hereditarianism (see below) to say that crime is genetic. But I don't think it's necessarily racist, because it doesn't specifically say anything about the presumptively natural groups of humans called "races."

Likewise, we sometimes hear that racism is about the exercise of social power, and that therefore poor people cannot be racists, because they lack social power.

But that is also unsatisfactory, for the same reason—it eliminates "race" from the concept of "racism." If racism is really about power, it seems to me that it should be called "powerism." And furthermore, there is something a bit unsettling about turning a blind eye to hateful poor people. Can't they be just as racist as rich people? Although they lack the institutional means to implement their hatreds, history shows that angry mobs can be awfully destructive every now and then. Why should they be absolved of responsibility for their hatreds simply on account of their lower socioeconomic status?

Racism is, somehow, about race; but it's different from the folk-hereditary idea that all people can be naturally clustered into a few big groups. Racism is a folk-hereditary fallacy that presupposes the existence of such groups and judges individuals by recourse to the properties of the groups to which they are assigned.

It's bad science because it is anti-empirical—you don't need to learn whether any specific black man is stupid, because you know ahead of time that black people are stupid; you don't need to know whether any specific Jew is cheap, because you know ahead of time that Jews are cheap. And it's politically abhorrent, because it is a fundamental principle of our modern democracy that people should be judged as individuals and not by properties attributed to their groups.

Hence the obvious analogy to sexism, where the categories are more natural, but the fallacy is just the same.

Racism is a folk-hereditary fallacy because its central presumption is that the attributes of the race have inscribed themselves on the constitution of the member.

It is important to note that the fact that races don't exist as natural entities has no bearing on racism. Racism exists as a real, social fact. In other words, racism directed against Jews, Irish, and Puerto Ricans is no less racism by reason of the fact that its victims are not members of natural biological units.

That is less of a paradox than it may seem. Christianity exists as a social fact, whether or not God exists as a natural fact; and of course the workings of the Church in shaping the modern world have been far more obvious than those of God. Social facts can be immensely powerful forces regardless of whether there is a basis in nature for them.

The major paradox of twentieth-century anthropology is that it arrogated to itself the definition of what the "real" races are, particularly with the accession of the Nazis in Germany. Physical anthropologists in America were at great pains to denounce Nazi policies and ideologies, focusing particularly on their "racial" isolation and persecution of Jews and Gypsies. American physical anthropologists criticized the Germans by denying that Jews and Gypsies were "really" races; for only a true expert knew what a race "really" was.

Well-intentioned as the argument was, it was like medieval scholastics debating about angels. The issue wasn't race, it was racism. People who hate Jews do so independently of the pronouncements of scientists (although scientists can lend tragic credibility to their bias), and independently of the facts of genetics (although natural differences can appear to lend credibility as well). Racism exists with or without science to back it up. Scientific racism is abhorrent to those who wish science to be benign and liberating, or at least neutral. But to see genetics invoked for oppressive ends taints science with the hue of demagoguery. So dewy-eyed geneticists sometimes tell us that genetics will undermine racism by showing that races don't exist.

Dream on.

The problem, as a molecular anthropological perspective can reveal, is that group hatreds are not genetic but folk-hereditary issues. For what biological property can indelibly inscribe itself upon the core of every member of one group and upon no members of another? People fight over many things, and frequently they vilify and demonize their opposition in order to win. To the extent that the opposition consists of groups of people, those people become objects of hatred because they embody the nature of the dispute.

But the embodiment is symbolic, not biological.

So genetics won't go very far toward solving the problem of racism, because racism hasn't much to do with the science of genetics. It existed long before genetics.

There's another aspect to the linkage of genetics and racism. I'm always astonished to find it asserted in the sociobiological literature that humans have a deep hereditary propensity for "xenophobia," fear or hatred of others, or more grandiosely, a genetic basis for genocide. Now of course, one needs to tread a delicate line here. One does not want to celebrate or glorify martyrdom and victimization, for that serves to parochialize the experience and limit its meaning. Genocide is an enormous tragedy and acknowledging its universality helps to promote the empathy that may discourage future holocausts. But what are the implications of saying that it's genetically based? After all, in acknowledging its universality, one does not want to trivialize genocide either.

The argument presented in sociobiology, and often presented as science, is how easy it is to hate and want to kill others unlike you. So easy that it's virtually universal. "Xenophobia" appears to be widespread in the human species. It knows no racial boundaries. So everyone has the taint of evil about them; no group is immune to it. But exactly what constitutes the fear or hatred of "others"? How do you actually know who the "others" are? How can you recognize them when you encounter them? What really are those people "unlike" you, and how have they come to be unlike you?

It is not, of course, reasonable at all to suppose that group antagonisms occur only between genetically or biologically different peoples. Were the Hatfields and the McCoys biologically distinct from each other? Or the Viet Cong and the South Vietnamese? Or the Bloods and the Crips? And that's precisely the central point: xenophobia, whatever it may be—or more to the point, the perception of "otherness," of alienness—is not based on natural differences. It's based on language, the deity worshipped, traditions, diet, activities, beliefs—things that are learned, not things that are innate.

Alienness is thus a construction, not a fact of nature. Whom you perceive as foreign, or who is to be suspected or hated or even worthy of death, is generally based on cultural traits and on cultural histories. The greatest genocidal hatreds are between peoples who are biologically very similar: Hutu and Tutsi, Bosnians and Serbs, Israelis and Palestinians, Huron and Iroquois, Germans and Jews, English and Irish.

That is why I think the argument for the potential universality, and hence the genetic basis, of genocide is a biologically trivial one, because it presupposes a natural difference between the two groups, the oppressors and the victims, which often does not exist. How, then, can it possibly be important in explaining or understanding the meaning of genocide?

Whom your group hates is defined culturally. Why your group hates them is defined culturally. What you're expected to do about it is defined culturally. There is no merit that I can see in talking about any biological basis for genocide, because such a basis, if it even existed in the first place, accounts for nothing.

The lesson of the Holocaust, therefore, lies not so much in the attempt on the part of one group to destroy another, which is indeed a recurrent tragic theme of human global history, but rather in the recognition that it was carried out by Europeans against themselves, and that it took place in an age in which some form of enlightenment was thought to have existed.

What we gain from presupposing genes for genocide is unclear. All this serves to do is to absolve the guilty of responsibility, because "It wasn't our fault, it was just human nature," which is certainly a perverse use of genetics.

So to the extent that we can establish that racism is not biologically significant, the genetic basis for it is meaningless.

More to the point, it is folk heredity.

But the fallacy of racism does not only make itself felt in the form of genocide. The central fallacy lies in dehumanizing a person because you perceive them as a member of a group and not as an individual. This cuts both ways.

I was recently asked to review a book about an issue the author melodramatically regards as the subject of a conspiracy of silence, for it is apparently taboo—the genetic superiority of blacks in sports.

Now, everybody knows that blacks are prominent in major sports. Look at basketball, look at defensive secondaries and running backs in football, look at boxing. But how do you get from "prominent" to "racially superior"? There are three major sets of variables at work in making someone prominent in athletics: individual aptitudes (whatever they might be), social and cultural setting (including expectations, opportunities, what is considered a respectable occupation, and the like), and any group endowments. With three causes and one effect, you can't reasonably draw a conclusion about the origin of black superiority in sports.

Thus, there are rather few possibilities for someone to conclude that racial superiorities are at the heart of the observed prominence of blacks in sports. Either the speaker is incapable of rigorously drawing conclusions from data or else the racist conclusion preceded, and is independent of, the data.

The author indignantly denies being either illogical or a racist. He insists he wasn't saying blacks are innately worse than whites, as for example Herrnstein and Murray did for intelligence in *The Bell Curve* (which he took pains to repudiate), but rather, that they're innately better.

But whether you think "they" are better or worse is a trivial difference, for the fallacy lies in attributing any innate property at all to "them" in the absence of genetic evidence.

The issue is, what constitutes scientific evidence, and a rigorous scientific argument, for innate group-level differences? You can point to Afro-Caribbean sprinters, or Kenyan marathoners, or Michael Jordan and the NBA until the Second Coming, and it won't say anything about the genetic propensities of black people in athletics, any more than the brilliance of Jack Benny and the predominance of Jews in comedy indicates the presence of comedy genes in Jews. The fallacy of *The Bell Curve* is not that it focuses on the supposed racial basis and innateness of an undesirable trait—stupidity—but that it infers the innateness of the trait simply from observation of predominance (of the other races). In fact, the athletic argument is essentially identical to *The Bell Curve*'s: Look, there's such an obvious and consistent pattern of this trait in those people; it must be innate; Q.E.D.

The fact that something is consistently observed does not imply that it has a genetic cause. We know that. If you want to argue about science and about genetics, you need controlled data and genetic data.

The author demands a refutation, calling me (and placing me in some pretty fine company) an "environmentalist," to make it sound as if I were a tofu-eating tree-hugger. But as we noted earlier, it is a basic tenet of science that the burden of proof always falls upon the claimant. I don't have to refute the claim that blacks are racially superior as athletes, any more than I have to refute the claim that angels cause mutations, or that stepping on a sidewalk crack breaks your mother's back. If you want to participate in a scientific discourse, you have to abide by its rules, or you'll end up shaking your fist at the sky and muttering, "The fools! They called me mad! But I'll show them. . . . I'll show them all!"

What would it take to demonstrate that blacks are innately gifted as athletes by reason of race? Above all, it would require delimiting the problem in a far more precise manner and confronting the complications surrounding the answer. It would involve acknowledging the complexities of life histories, the lack of rigorous data on the subject, the ease with which cultural stereotypes can be made to look like natural differences, and the difficulty in generalizing about the properties of populations from a comparison of the performances of their most outstanding members. And that's just for starters. Otherwise, you present a lot of interesting fluff that doesn't prove anything about anybody's innate abilities, much less about racial differences.

It would require acknowledging that black athletes are physically quite diverse; certainly, a black interior lineman in football is physically very different from a black point guard in basketball. Thus, the innate gifts they each possess that enable them to earn a living at professional sports are probably not the same ones—and are probably more likely explicable as individual, not racial, gifts. The racial issue is more likely why a black person is more apt to see the exceedingly high-risk world of athletics as a reasonable venue for earning a livelihood than a white person is.

Darwin's Athletes by John Hoberman eloquently presents the case for *dis*believing that black predomination in sports is caused by constitutional factors of the race. It's not that we can say specifically why one person becomes better at something than another person—that's fodder for astrologers, not scientists—but rather that there are simply a lot of forces at work besides imagined racial propensities.

Here the fallacy of racism has been inverted, but it is a fallacy nonetheless. It's not whether blacks are innately better than, or worse than, whites at something. Rather, it's that there is no reason at all to assume that any observation of difference is due to a trait that is either innate or racial. How easily, for example, Muhammad Ali and Michael Jordan can be made to slide from being *extraordinary, well-trained* black men to being *representative* black men.

And that's precisely the problem. Athletes are elite performers and aren't representative of anyone but themselves, except symbolically. So what can it mean about "black people" if the ten fastest known sprinters are all black? In the first place, it is a gross perversion of statistical sensibilities to characterize a population by its ten most extreme members. And in the second place, the fastest white and yellow people are not too far behind anyway—we're only talking about the twinkling of an eye here, after all. And how do you know you haven't missed some really fast white guy somewhere?

Finally, those who laud the innate aptitudes of blacks for basketball frequently need to be reminded of the push a decade ago to allow American professionals to play in the Olympics. The rest of the world, it seemed, had caught up to our gifted, predominantly black, amateurs.

HEREDITARIANISM

The Bell Curve began with a premise that is not racist: that in general, differences in IQ, or in performance of some surrogate pencil-and-paper test, are constitutional, or innate.

This does not presuppose the existence of races or of group-level differences (although its authors assume those too). To the authors of *The Bell Curve*, being a doctor or a lawyer or a professor means that you are simply likely to be smarter than a carpenter, fireman, or television repairman. It's an assertion as much about innate intellectual differences *within* groups as between them.

Although the distinction between differences that may exist *within* groups and those that may exist *between* groups may sound trivial, it isn't. As already noted, most behavioral variation in the human species is *between* groups, yet most genetic variation is *within* groups. That makes it exceedingly unlikely that the latter can be a major cause of the former.

Perhaps the weirdest weapon in the hereditarian arsenal is the concept of "heritability," borrowed from animal and plant husbandry. Technically defined as the ratio of genetic variation associated with a particular feature to the total observable variation in that feature, and varying from 0 to 1, heritability is hard to measure in people. Consequently, we rely on a number of shortcuts. The problem with heritability is that *it sounds like a property of the feature itself, when in fact it is merely a description of the population in which the trait appears.*

The classic example is to envision two plots of soil, in each of which a handful of seed is planted. The plot on the left receives ample water and fertilizer; the plot on the right doesn't. Plants in the left plot grow to be tall and vigorous; since they are not all identical, but their environment is largely homogeneous, most of the variation in height is due to the genetic differences among the seeds originally planted. Thus the heritability of plant height here is quite high. Plants in the right plot are small and stunted; but likewise the environment is homogeneous, so differences in height are largely due to genetic differences, and heritability of height is high. *But the large difference in average height between the two populations is due entirely to the environmental difference imposed upon them.*

Thus, heritability is a description of a population, not a property of the trait. From our experiment in heritability above, we learn little about "height"; rather, we learn something specific about each particular plot of plants.

When you hear someone say that the heritability of intelligence is 0.4, for example, the correct question to ask is, "In whom?" This is a statement about a particular group of people, not about intelligence. Moreover, however genetic it sounds, it is culturally loaded, and its meanings and usages aren't derived from the technical sense it has in the science of genetics.

Hard as it may be to believe, we have actually learned something about human heredity and human nonheredity in the past hundred years. Most important, we have come to realize that when we ask, "Is a feature genetic?" we are really asking, "Is the detectable variation [in the feature] due to genetic causes or to other causes?"

We have learned that a consistent observation of difference between two groups does *not* necessarily imply a genetic basis for that difference.

A glance through any of the major science journals will suffice to show that contemporary genetics is characterized by a particular language, mode of argumentation, and evidentiary standard. The simple belief that crude hereditary factors are important in human life certainly long antedates the science of genetics and is therefore independent of it, but it often piggybacks on the credibility of genetics. Such beliefs are cultural, and therefore widely pervasive, even among geneticists, but they are different from scientific inferences about genetics; that is, hereditarian beliefs are distinct from science. The failure to make this distinction is retrospectively a crucial error of the 1920s—geneticists at that time widely felt that being a geneticist implied a commitment to constitutional, hereditary explanations for human social differences, even when they lacked valid evidence for such explanations.

Scientists today regrettably sometimes fall into the same trap. "We used to think our fate was in the stars. Now we know, in large measure, our fate is in our genes," the molecular geneticist James Watson told *Time* in 1989. One may question the existence of fate, its localization to our cellular nuclei, and whether genetics is indeed at root merely high-tech astrology (though presumably more accurate). In translation, of course, the statement is nothing but a grant proposal—the assertion about the centrality of genetics to everyday life was a sales pitch for the Human Genome Project. This helps highlight, however, the conflict of interest still faced by geneticists in promoting the importance of their field, for their self-promotion expresses *both* a funding request and a social philosophy. The crux of the matter—as the lessons of the 1920s show us—lies in distinguishing the ideas of modern genetics (i.e., contemporary science) from folk ideas about heredity, even when those ideas are held or promoted by geneticists.

Sometimes we encounter modes of thought inherited almost directly from Charles Davenport and the genetics of the 1920s: genes encode the development of the brain, which is the seat of the mind, which is composed of thoughts, which lead to deeds. Therefore bad thoughts and acts are caused by bad genes. From a bully pulpit as editor of the journal *Science,* Daniel Koshland proclaimed the importance of the brain and falsely promoted the ability to infer organic defects from the observation of bad deeds; a few years later, he decried a German court's leniency toward Günther Parche, assailant of the tennis star Monica Seles, on the basis of its ignorance of brain disease. Yet Parche was never diagnosed as having a brain disease—much less any kind of constitutional cranial defect—some faulty "wiring" that led him to the act. Would the criteria for an adequate knowledge of modern science include not making a diagnosis without examining the patient?

Unfortunately, scientific zeal comes with a responsibility borne of the authority of science and of scientific statements. This is the legacy of the genetics of the 1920s, and it is one that the current generation must be better educated about and better prepared to assume. It cannot be up to outsiders to monitor the pronouncements of geneticists; it must be up to geneticists themselves to distinguish between their authoritative statements as scientists and their idiosyncratic or culturally produced ideas as citizens.

Some of the most persuasive folk knowledge comes from anecdotes about identical twins. In the world of twin studies the unscrupulous and the credulous symbiotically plumb the depths of contemporary pseudoscience.

In one recent popular book, called *Twins: And What They Tell Us About Who We Are,* journalist Lawrence Wright introduces the reader to a psychologist at the University of Minnesota, Professor Thomas Bouchard. Bouchard, Wright tells us, had long been teaching the controversial ideas of Arthur Jensen and Richard Herrnstein to the effect that IQ differences are due to genetics, especially IQ differences between races. Then Bouchard discovered the Jim twins.

Who are the Jim twins? Identical twins born in 1939 in Ohio, who were reunited decades later and found to have: the same first name, first wives named Linda, second wives named Ann, dogs named Toy, and sons with the same name.*

"The reunited twins story is a veritable chestnut in journalism," Wright says, but no red flags go up for him.

Bouchard begins to study the Jim twins, particularly their performances on pencil-and-paper tests, and finds them to be amazingly concordant. Then, with grants from the Pioneer Fund, he begins to study other twins as well. What is the Pioneer Fund? Wright obliges us: "a New York foundation that has roots in the eugenics movement of the thirties and that has a history of backing projects that advocate racial separatism. The Pioneer Fund has given [Bouchard's] project over $1.3 million, more than any other project in the fund's history."

*We learn more about them in a picture book called *Twins* (Philadelphia: Running Press, 1998), containing essays by Ruth and Rachel Sandweiss and photographs by David Fields. One was told he had a brother, but unaccountably did not try to find him until middle age.

> When they were young, living only forty-five minutes apart, they both grew up with adoptive brothers named Larry.
> They smoked the same brand of cigarettes, drank the same brand of beer, . . . and enjoyed math and disliked spelling in school. They used the same slang words. . . .
> "We found out that we even vacationed in the same spot on the Florida coast,"
> Jim Springer adds, "at about the same time of year."

The curious reader might be led to inquire just what the Pioneer Fund sees in it.

But still no red flags go up. "There has simply been nothing on the environmental side," the journalist observes, "to counter the power of twin and adoption studies."

But what power? Let's think about this. A committed ideologue scientist, with funding from a radical organization (which would achieve greater notoriety for their funding of much of the racist work cited in *The Bell Curve*), builds a research program on patently idiotic stories of reunited twins, which should be of greater interest to mythologists than to geneticists. And then he promotes it to the media as evidence for Nature 1, Nurture 0—and the journalist doesn't find a reason to be skeptical?

A bit of reflection permits the realization that the Jim twins are quite hard to explain scientifically. There are in fact only a small number of possibilities available to explain the story. What are they?

First, perhaps it's just an odd coincidence, like the list of amazing, dopey similarities between presidents Lincoln and Kennedy—assassinated 100 years apart, one in Ford's Theater, the other in a Lincoln Ford, both succeeded by Johnsons, and so on. In which case, it isn't worth mentioning. Coincidences are of no interest to science. On the one hand, any two people at random looking for similarities between themselves will inevitably find them. On the other hand, it does boggle the mind that two men married to two identically named women, with identically named sons and dogs, and with the same name themselves—an extraordinary set of occurrences in and of itself—would then turn out to be identical twins separated at birth! The mere fact that it is invoked in print implies it is not intended to be regarded as a coincidence.

So let's move on to the second possibility. Maybe these amazing similarities indeed attest to the genetic unity of identical twins. In fact, I raised that question during a plenary talk I gave in 1996 to the International Congress of Human Genetics. I said, "You people are an audience of human geneticists, and I would like to know, How many of you think that the name you give your dog is under some kind—*any* kind—of genetic influence?" I can happily report that not a single hand went up. The blunt fact is, there isn't a competent geneticist in the civilized world who will look you in the eye and tell you that the name you give your dog is under any form, however cryptic, of genetic influence.

That, of course, doesn't prove that there is *no* genetic cause. But it does show that the people who understand the most about genetic evidence, and are the most critical thinkers about genetic issues in the human species, do not agree with the weird psychologists who present the Jim twins as presumptive evidence of genetic control. The anecdote about the Jim twins actually demonstrates *nothing whatsoever* about the genetic unity of identical twins.

Third, then perhaps it demonstrates the psychic ESP bond of twins. Even our wide-eyed reporter, Lawrence Wright, hesitates here: "Clairvoyance is a part of twin lore; . . . [t]hese suggestive psychic connections between identical twins could explain some of the mysterious synchronicities, but they have been rarely tested and never confirmed." But this actually was a straightforward subheading in *Newsweek*'s cover story of November 23, 1987, on the amazing separated-twin research, which included the Jim twins. (And why not? Most readers of *Newsweek* believe as strongly in ESP as in genetics—twin studies are a strong site of convergence between folk knowledge and science.)

But wait a minute—these people are not being studied for psychic powers; they're being studied for genetics and personality. We are being asked to judge this as scientifically competent data. How on earth did the dubious end of behavioral genetics come to be supported by the pseudoscience of parapsychology?

After all, if the twins are in psychic contact, then there is an immediately detectable flaw in the majority of the data Bouchard's study collected, which are the results of pencil-and-paper tests, responses to the same questions about personality and mental processes, which were subsequently compared for concordances.

It is a fundamental assumption of any such situation that the people writing down their answers are not in psychic contact with each other. Those test concordances are now invalid because they are subject to an uncontrolled variable—if there is a real possibility that the twins may be trading answers psychically.

Imagine the chaos that would ensue if it were known that identical twins taking the SATs were cheating by using ESP!

But of course, there is no ESP. These twins don't have it, nor does anyone else. ESP isn't science, and therefore doesn't constitute a reasonable scientific explanation for anything, because it lies outside the domain of science. As an explanation for similarities of twins, it is useless and inane. Worse, the more closely behavioral genetics is linked to ESP, the less credible it necessarily becomes.

The process of elimination brings us to the fourth possible explanation.

There is just simply something screwy with these stories.

And this is explained by another simple conflict of interest: identical twins *without* amazing coincidences and psychic bonds don't get written up. (As one of my undergraduate students, who is an identical twin and *not* in psychic contact with her sister, said to me: "I guess we'll never be on Oprah.") What we encounter here is generally a strange brew of science and pseudoscience; of rigor and pop fantasy. Unlike scientific data, such "gee-whiz" stories are often told, but rarely rigorously analyzed, and never confirmed.

Here is an explanation for the similarities of identical twins. They are similar because they are genetically identical. They are also similar because they are usually raised to emphasize their similarities. When raised apart, they are often placed in similar homes—adoptive families, regretfully from the standpoint of experimental design, are very homogeneous. *And the most bizarre stories are just not true, exaggerating either their concordances, or their separateness, or both.*

Opposing the amazing twin anecdotes, there is a community of counteranecdotes developing. The journalist Lawrence Wright reports that the Jim twins had known of each other and had met each other on many occasions prior to their "reunion." (Oddly, he finds that less credible than their amazing coincidences!)

Another amazing set of twins are Jack and Oskar, born in 1933 and separated in the divorce of their parents; one grew up Jewish in Trinidad and the other in a Nazi home in Munich. They both, Constance Holden writes in *Science,* "think it's funny to sneeze in a crowd of strangers, flush the toilet before using it, store rubber bands on their wrists, read magazines from back to front, dip buttered toast in their coffee." But they each knew of the other and had met several times as adults, and their families had kept up a correspondence. And the counteranecdote is that the Jewish twin, who now lives in California, appeared on talk shows on diverse topics with such frequency that he joined the actors' union, and that the twins were attempting to sell their life stories to Hollywood.

Hardly the stuff of innocent birth-separation-reunion scenarios. Much less of scientific inferences about genetics.

ESSENTIALISM

The fourth major fallacy of folk heredity is that old philosophical Platonic problem known as essentialism, which we encountered earlier in our discussion of race. As a scientific fallacy, it refers to imagining underlying uniformity in the face of apparent diversity.

Consider, for example, the question of human nature. Humans do, and have done, an extraordinary number of things as individuals and as collectivities. Presumably, these are all reflections of human nature, some genetic program distinct from cow nature or dog nature or chimpanzee nature. It is human nature, we can safely assert, to walk on two legs.

But is it human nature to have an extramarital affair?

Is it human nature to go to war?

Each of these is a folk-hereditary statement about our underlying makeup, our basic constitutions. If they were statements from the realm of genetics, we'd say "our genome." But they aren't statements from genetics, based on genetic data. They're statements from some other intellectual domain, known these days as "evolutionary psychology."

Evolutionary psychology dons the mantle of Darwin to tell us why various components of our behavioral and mental repertoire evolved. Unfortunately, it hasn't shown itself to be particularly good at (1) identifying the fundamental components of our behavioral and mental repertoire; (2) establishing their evolutionary histories; or (3) proving that their inferences reflect an underlying human nature, rather than being simply the projection of contemporary social values.

Take a well-publicized example: the waist-to-hips ratio of women. If you draw silhouettes of women's figures and ask American college students which ones they think are most attractive, they will invariably go for someone shaped like Marilyn Monroe. In fact, if you ask the same question of many men around the world, they will generally also go for Marilyn Monroe. Aha, say the evolutionary psychologists, a clear indicator of a fundamental property of the human mind, which has evolved over the eons.

But is it? Are American college students valid representatives of the primordial human mind, or are their beliefs, thoughts and preferences molded by both their humanness and their twentieth-century Americanness? Does the fact that other men around the world find Marilyn Monroe's figure attractive mean that they are expressing a basic human propensity or merely that their sexual preferences have been influenced by the American entertainment media?

After all, American tastes and preferences have been globalized. The world's cultural diversity is constantly shrinking as a result of the relentless economic and social pressure of the United States. Not that it's necessarily a bad thing; only that with increasing uniformity of thought, it becomes ever more difficult to distinguish the properties of the human mind from the consequences of human social history. It is an old anthropological saw that what you do seems natural; how much more natural it is to see the whole world do it! But what is widespread is not necessarily innate—that's the fallacy of essentialism.

There is a simple test one could perform to try to judge whether the widespread attraction for men specifically of Marilyn Monroe's shape is an evolved, innate propensity or an artifact of twentieth-century American taste disseminated to the rest of the world. You could test a remote group of people unfamiliar with Marilyn Monroe, or American movie stars and pinups generally, and see what women's figures they prefer. The problem is, of course, that there is hardly anyone these days who falls into that category—you can see the latest Julia Roberts movie in Nairobi.

But when biologist Douglas Yu and anthropologist Glenn Shepard visited the Machiguenga in the South American jungle, they had precisely that opportunity—to test the idea that men naturally prefer women shaped like modern movie stars. And what did they find? That the Machiguenga men preferred women shaped the way their women were shaped and not the way American movie stars and models are shaped.

The discovery may evoke a schoolyard reaction of "No shit, Sherlock!" but it is an important one, because the possibility of performing such studies is rapidly diminishing. That will make it ever more difficult to refute essentialized, folk-hereditary assertions about human nature.

Consider another widely disseminated result. An evolutionary psychologist sends questionnaires out to people in thirty-seven different countries asking what men and women are attracted to as properties in mates. He finds that in general, women want a sugar daddy and men want a babe. And that must be the way we evolved, argues David Buss, author of *The Evolution of Desire*. It's natural, it's evolved, it's genetic.

And that ostensibly scientific result enters the lore of evolutionary psychology, repeated in popular scientific works with titles like *The Moral Animal* and *The Anatomy of Love* and *Consilience*.

But wait. How different are people in those thirty-seven different countries? The French and Swiss and Italians may be different, but they share many common cultural values as a result of their history. Add that complication of the American media's impact on the rest of the world, and suddenly it becomes clear that those thirty-seven data points are neither independent of one another nor an impressive representation of the breadth of human thoughts, actions, and capacities.

Even more significant, however, is the fact that patriarchal values are widely pervasive in agricultural and industrial societies. Women have variable, but generally lower, status than men. Even in America, a woman typically earns only about three-quarters as much as a man in an equivalent position. The only sure strategy to gain access to financial resources where men have preferential access to them is to marry them. Is the fact that women worldwide express preferences for sugar daddies, therefore, an expression of innate human nature— or just an obvious rational response to a social problem faced by women worldwide?

Unless we can distinguish between those two possibilities—and these scientists don't—then the data are valueless, for they don't permit us to conclude that it is in fact human nature that we are seeing. It could just as easily be a response to human social history. And the fact is, as Cornell anthropologist Meredith Small points out, there are enough exceptions out there to make generalizations about human nature exceedingly tenuous. What about those exceptions? Are they mutants? No, they constitute the crucial evidence against the idea that there is a fundamental human nature reflected in the crude generalizations detectable in human thoughts and deeds. That concept of human nature is an essentialized fallacy.

Cultural globalization imposes a uniformity of thought on the world that will make it ever more difficult to detect exceptions; but it is precisely those exceptions that reveal the fallacy of essentialism, of attributing to an underlying, invariant, and imaginary human nature the products of a specific, historically situated culture. And consequently, when Northwestern psychologist Alice Eagly controlled the data on those thirty-seven different countries' mate preferences for economic inequality by sex, those patterns quickly vanished.

Studies of the genetics of homosexuals, as noted earlier in a different context, frequently begin with an assumption of essentialism—that "homosexual" is a noun describing a kind of person, rather than an adjective describing a kind of act. How many homosexual acts does it take to make someone "a" homosexual? That usage of the word is in fact only a product of the past few centuries.

While there are many people today who have a sexual identity as "a homosexual," it is clear that the construction of identity is partly a product of the available categories, the "kinds" of people available to be. That is cultural knowledge, not biological data. From the perspective of anthropology, what is striking is the manner in which the action has been essentialized into the constitution of the actor through history; once it is acknowledged culturally as a state of being, it is then adopted (or imposed) as an identity of individual people; and once it is perceived as an identity (specifically a deviant identity), it is then sought in the genes. From the perspective of molecular anthropology, the whole endeavor reflects a folk-hereditary fallacy; there's no reason to think it was ever in the genes in the first place.

RESPONSIBILITY AND FOLK HEREDITY

It is the responsibility of a scientific community to distinguish for everyone else the science from the pseudoscience: promote the former and debunk the latter. The geneticists of the 1920s were either unwilling or unable to do that. The geneticists of the twenty-first century have to learn from their predecessors' mistakes. Behavioral genetics as a community has to meet the obligation of skepticism—it cannot be up to others to question the validity of the work; science has to be rigorously evaluated from within.

Behavioral genetics has to be done in the context not merely of scientific knowledge but of *humanistic* knowledge. We are not starting at ground zero here. Without the benefits of confronting the class issues and history surrounding the genetics of criminality, or the race and culture issues around the genetics of intelligence, it cannot be good scholarship. And if it isn't good scholarship, it cannot be good science. That is the burden imposed just by tackling tough and important questions.

What we need to know is whether specific behavioral genetic studies reflect modern genetic sensibilities or merely hi-tech folk wisdom about heredity. As I noted at the outset, contemporary genetics relies on the ability to relate an observed characteristic (a phenotype) to a hereditary state (a genotype) causally and mechanistically. Studying phenotypes alone, or their distributions, or associations among them, and then blithely inferring genes for them, is simply inadequate in the context of modern genetic reasoning. To invent genes where none are known is poor genetic practice, with a long and sordid history; it confuses fantasy with modern science. While there may be short-term benefits to some scientists who pursue these lines of argument, they are counteracted by the long-term harm both to the science of genetics and to the authority of science generally.

And it is the responsibility of behavioral geneticists to point this out. If they don't, the rest of us can't be expected to.

EPILOGUE

The millennial issue of the *Scientific American* includes an ostensibly balanced and objective article called "The End of Nature versus Nurture" by Frans de Waal, a primatologist. While sensibly eschewing "simple-minded genetic determinism," he also reports that "the evidence for a connection between genes and behavior is mounting. Studies of twins reared apart have reached the status of common knowledge. . . ."

That is, of course, the problem. They are *nothing but* common knowledge. De Waal's *Scientific American* article is actually illustrated up front with a set of identical twins, and tells us, "Twins reared apart have been studied for clues about the relative contributions of genes and environment to human behavior. These brothers," it goes on, "rediscovered each other later in life when both were mustachioed firefighters."

So genes determine who wears a moustache?

Or genes control being a firefighter?

Otherwise, what is the point of calling this to our attention?

Seven

HUMAN NATURE

IN A RECENT BOOK CALLED *Demonic Males: Apes and the Origins of Human Violence,* primate sociobiologist Richard Wrangham and his co-author Dale Peterson write that it is a natural inclination of the human male to be aggressive—to be "demonic." And this inclination, they tell us, "is written in the molecular chemistry of DNA."

Even for the moment overlooking the crude essentialism of the thought, the question that jumped out at me when I was asked to review the book for the journal *Human Biology* was: What exactly is the evidence from the molecular chemistry of DNA?

In fact, this wasn't a book about molecular chemistry, nor was any significant chunk of the book devoted to the subject. There was no evidence provided for a "demonic" gene on the short arm of chromosome 5, nor on any part of any chromosome. That bit about molecular chemistry of DNA was simply a purple way of expressing the premodern thought that men are just innately nasty. They might just as well innately be anything you like, of course, because if genetic evidence is unnecessary, you can make genetic assertions with impunity.

This particular argument for the innateness of male demonism is based on the conjunction of two prior arguments: first, that chimpanzee males are intrinsically demonic; and, second, that humans are genetically almost identical to chimpanzees. Therefore, human males must naturally be innately demonic as well.

The latter part of the argument has already been dealt with. In spite of the genetic similarity, we are not chimpanzees and are easily distinguishable from them. We're the ones walking upright, speaking, weeping, laughing, praising, insulting; we're the ones with erotic sexual foreplay and sex games, the ones who fall in love, who cook our food, who decorate ourselves for public display. So different from chimpanzees, and yet all the while genetically so similar to them. Whether or not chimpanzees are demonic, there is plenty of genetic leeway for humans to have become effectively angelic since we parted gene pools with them.

And yet the behavior of chimpanzees is not entirely irrelevant to understanding ourselves. We are derived from a recent common, if nebulous, genetic substrate. If chimpanzees really are fundamentally demonic, or angelic, or [*insert your own adjective here*], that might have some relevance to our understanding of humans. It might represent the ancestral form of human thought and behavior, something we have either retained or overlaid (like the urge to explore and touch, color vision, and mobile shoulder of chimpanzees). On the other hand, it might represent something we once had but have since altered or sloughed off. Like the large canine teeth, body hair, and grasping feet of chimpanzees.

Unless, of course, it simply represents a unique evolutionary development specifically of the chimpanzees themselves—like the stripes of a zebra, the blue nose of a mandrill, the call of the howler monkey, the moustache of the moustached tamarin, or the beard of the bearded saki.

APES AND PEOPLE

There is consequently no guarantee, short of detailed physiological and genetic data and analyses, that anything chimpanzees do is directly relevant to understanding anything that humans do. Since they have been different species for several million years, anything that chimpanzees do may be either (1) an element shared with human nature; or (2) an ancient element of human nature now lost by humans; or (3) an evolved element of chimpanzee nature, never possessed by human ancestors.

The opposite idea, that what an ape does is illuminating for human nature, has come to be known as "the naturalistic fallacy." Let us say, along with the nineteenth-century mutton-chopped English polymath Herbert Spencer, that the natural world is governed by unfettered competition. Does it not follow that competition, unrestrained by governmental intervention in the form of poor laws, or welfare, or social safety nets, is the most appropriate and natural form of human society?

On the contrary, argued Spencer's friend, the great biologist Thomas Huxley, who had actually published the first book on human evolution, *Man's Place in Nature,* in 1863. Thirty years later, Huxley had come to realize the philosophical abuse that biology could wreak on society. And in an essay called "Evolution and Ethics," he challenged Spencer's worldview: "Let us understand, once and for all," he avowed, "that the ethical progress of society depends, not on imitating the cosmic process, still less in running away from it, but in combating it." Humans do not live *in* a state of nature, they have *risen from it.* What goes on out there among the birds and the beasts may have no bearing at all on what is natural, much less on what is appropriate, for humans.

Apes go around naked and sleep in the trees, and yet humans do not. Which condition should be considered more natural for humans—what apes do and humans do not (that's a perverse conception of human nature!); or what humans do and apes do not (in which case, why bring up the apes in the first place?)?

What bearing, then, do any data from apes have on our understanding of humans? Obviously, we must look for commonalities between apes and humans.

CULTURAL PROJECTIONS

In the 1960s, baboons were taken to be the most valuable models available for understanding the evolution of human behavior. Why baboons? Their closest relatives, the vervet monkeys, are all highly arboreal, but baboons have descended to the ground for most of their daily routine.

Rather like humans.

Moreover, the males vie with one another rather conspicuously. They grunt, posture, bark, chase, and nip—all with the obvious goal of intimidating and displacing or overcoming all rivals. That seems to be the core of their society.

Also rather like their human counterparts.

But in one key way, it was believed, they were not like humans.

They didn't hunt.

So hunting must have made all the difference in human evolution. Here you had a species not terribly distantly related from us, with similar hands, feet, teeth, and brains; recently descended from the trees, like us; socially rather like us; and had they only, perhaps, developed the ability to hunt—what we might call "the skill to kill"— they might have been us.

And yet, since the 1960s, that interpretation of baboon society and its relationship to human nature has fallen out of favor. First, baboons do indeed hunt, it turns out. They've been observed to chase down and eat several different kinds of animals—small monkeys, bush pigs, gazelles. Hunting doesn't contribute a major component of their diet, but it has been observed regularly enough to now be accepted as a facet of baboon life. So the big difference between us and them turns out not to be as profound as the theory requires it to be. If they hunt, after all, then they should be us—but they aren't.

Moreover, the male interactions—so conspicuous, so dangerous— turn out to be largely ephemeral. Long-term studies of the course of a baboon's life have revealed that baboon females grow, mature, and remain in their social group. Males, on the other hand, transfer into a new group as they mature. As a result, the composition of a baboon troop consists of related females who grew up together and "know" one another and males who did not grow up together. It shouldn't be a surprise, then, that the males fight and jockey conspicuously for social position, while the female interactions are more subtle. But the female bonds are also more pervasive and longer-lasting than the male bonds.

What seemed at first to be a reiteration of dominant themes of American society—male competition and aggression, female passivity—were more the projections of the primatologists than expressions of baboon nature. What's more, it had been assumed that all that conspicuous fighting that male baboons do could only be over one thing—female baboons. After all, what do men fight about, if not women?

But again, fieldwork in more recent years has shown that the connection between sex and aggressive competition in baboons is fairly weak. Baboon females have many strategies for choosing mates, and the biggest and strongest male isn't necessarily the one they pick. Baboon males, for one thing, are smart enough to form coalitions to ward off attacks from—and sometimes to depose—the single biggest male. So the boys don't really seem to be fighting just to show off for the girls, like American high school students—once again, a cultural projection onto the animals.

Baboon nature, it seems, has precious little to tell us about human nature; for what we derive from baboons turns out to be largely a function of what we project onto them. Baboons are interesting because they're baboons and fairly close relatives of ours. But their utility as models for understanding human biology hardly outlasted the Beatles.

Chimpanzees, however, are far more closely related to us than baboons. Their DNA is really, really similar to our own. Perhaps the key to human nature lies in them. And that's what modern students of chimpanzees say in their grant proposals: *You should fund my research because it will reveal secrets of human nature.*

But it won't, any more than baboon studies did.

The best evidence in support of that assertion comes from looking at the oeuvre of the foremost student of chimpanzees—one of the world's best-known scientists, Jane Goodall.

Goodall began studying chimpanzees in the wild at the Gombe Stream Reserve in Tanganyika (now Tanzania) in 1960, under the aegis of the charismatic paleontologist Louis Leakey. After nearly four decades of research, her *National Geographic* specials are fixtures on educational cable television, she's done a commercial for HBO, her speaking engagements fill the largest auditoriums on college campuses, and she remains an inspiration to young women entering science.

When Jane Goodall enters a room, she fills it with an almost celestial presence, a delicate and beautiful woman with a steel backbone, who has lived in difficult places as an alien human presence in the name of science for many years. Her work remains the longest continuous field study of chimpanzees, indeed of any primate species in the wild, and her works on that species are authoritative.

And yet they've evolved over the decades. They are more than mere reportage about what chimpanzees do. Goodall's books on the Gombe chimpanzees are not merely different from one another by virtue of the inclusion of more and more of the same kinds of data and observations. Rather, they incorporate new and different perspectives on the apes, perspectives that have changed with the decades.

Her first and most famous book, *In the Shadow of Man,* is a chronicle of Goodall's first decade at Gombe studying the chimpanzees. She succeeds brilliantly in conveying a sense of the life history of chimpanzees—birth, death, adolescence, sexuality, danger, friendship, family. She also manages to show a facet of "personality" for each of the Gombe chimpanzees—dominant Mike, popular Flo, irascible J. B., playful Gilka, clever Figan. These chimpanzees didn't have to worry about the Vietnam War, or inflation, or earning a living. They lived at peace with nature, in a society of conflicts and appeasements, of immediate gratification, and of sexual abandon.

In The Shadow of Man was published in 1971, and the chimpanzees of Gombe could be seen as a hippie commune.

By the early 1980s, however, the chimpanzees were being seen rather differently. They were now selfish strategists, manipulating others in pursuit of their evolutionary goals. They now also had a dark side: some were even occasionally infanticidal. Far from the integrated love fest of the 1960s, the Gombe chimpanzees were now more like the

Carrington family of television's *Dynasty*. They were now seen as considerably more violent than the first two decades of observation had indicated.

The 1990s brought a new vision to chimpanzee research. Studies at other sites, notably Taï in the Ivory Coast and Mahale in Tanzania, afford behavioral contrasts to many of the observations at Gombe. Multiculturalism thus became the order of the day in chimpanzee studies.

From a hippie commune to vicious and self-interested exploiters to multicultural pluralism—chimpanzee studies since the 1960s have reflected American changing social patterns. Obviously, this does not mean they are without value and that we know nothing about chimpanzees. In fact, it means quite the opposite: we have learned much about chimpanzees and much about studying and interpreting their behavior—and that what chimpanzees *do* is only *one component* of what goes into the study and interpretation of their behavior. The study of chimpanzee behavior turns out to be as much about ourselves as about chimpanzees, but not in the facile manner that relates chimpanzees directly to human nature.

Rather, it is about ourselves in the sense that we constantly contribute a cultural and symbolic framework to the interpretation of chimpanzees—and it is all too easy to recover that framework and mistake it for a contribution of the chimps, rather than for our own input.

In the case of deriving human nature from chimpanzees, it's clear that the chimpanzees are not *just* chimpanzees. They are symbolic of a past life and of a simpler existence. They are us, minus something. They are supposed to be our pure biology, unfettered by the trappings of civilization and its discontents. They are humans without humanity. They are nature without culture.

NATURAL MAN

Imagining people without culture has been a pastime of Western thinkers for millennia. Culture is clearly not innate; thus one is free to imagine what a group of people who lacked religion, buildings, fire, haircuts, or rules governing sexual conduct might be like. And the world is a big place—surely they must be out there somewhere.

Herodotus writes of cave dwellers in Africa who had no language and squeaked like bats. Pliny the Elder, who was so fascinated by the world's marvels that he tarried a bit too long admiring the eruption of Mount Vesuvius in 79 A.D., repeated Herodotus's brief discussion of the cave dwellers.

In the fascinating, and frequently silly, *Travels of Sir John Mandeville,* dating from the mid-1300s, one reads of an island called "Tracota, where the people are like animals lacking reason. They live in caves, for they do not have the intelligence to build houses; and when they see a stranger passing through the country, they run and hide in their caves. They eat snakes, and do not speak, but hiss to one another like adders."

But these people were, of course, imaginary. There was not much to be said about them, but that they lived somewhere and were brutes.

That situation changed with the discovery of the New World. Although there were obviously great civilizations in the Americas, there were also people living in an extraordinarily rude state (to the European mind) and speaking in utterly incomprehensible manners. As stories of them filtered back to Europe, the Indians became valuable to the newly emerging schools of political philosophy. The first to invoke them was Thomas Hobbes, in his monumental work *Leviathan* (1651).

Hobbes took a new and radical approach to political theory. He tried to justify his English monarchial law by contrasting it with the situation of "natural man." What would people be like, asked Hobbes, if stripped of the trappings of civility and law, relying only on themselves and their own acts for survival? The answer, reasoned Hobbes, would be a state of

> Warre, where every man is Enemy to every man; the same is consequent to the time, wherein men live without other security, than what their own strength, and their own invention shall furnish them withall. In such condition, there is no place for Industry;

because the fruit thereof is uncertain: and consequently no Culture
of the Earth; no Navigation, nor use of the commodities that
may be imported by Sea; no commodious Building; no instruments
of moving, and removing such things as require much force; no
Knowledge of the face of the Earth; no account of Time; no Arts;
no Letters; no Society; and which is worst of all, continuall
feare, and danger of violent death; And the life of man, solitary,
poore, nasty, brutish, and short.

And where might such a situation be found?

It may peradventure be thought, there was never such a time, nor
condition of warre as this; and I believe it was never generally
so, over all the world: but there are many places, where they live
so now. For the savage people in many places of *America,* except the
government of small Families, the concord whereof dependeth
on naturall lust, have no government at all; and live at this day in
that brutish manner. . . .

And thus Native Americans came to represent for European philosophers the idea of "man in a state of pure nature"—the beastly
"other" against whom the civilized state would form a contrast. Or
more to the point, a development, a rise—an *evolution.*

Hobbes's argument was immensely influential in European political
philosophy. The concept of the uncivilized, purely biological person
was also invoked by the generations that followed Hobbes. And their
favorite empirical grounding for this argument, the place where "natural man" could actually be found, and whose data could be invoked
for the purposes of polemic, was America.

John Locke, for example, challenged Hobbes's most basic point
about human nature being a state of perpetual war, yet supported his
own ideas in the same manner as Hobbes—by recourse to the lives
of savages living in a state of nature. He asks, for example, why money
would be necessary if the earth were bountiful and land were readily
available. "Thus," wrote Locke in his *Second Treatise on Civil Government* (1690), "in the beginning all the world was America, and

more so than that is now; for no such thing as money was any where known."

But the question of human nature, and the Indian as a means of revealing it, were nowhere more forcefully or clearly articulated than in the work of Jean-Jacques Rousseau. He asks at the beginning of his *Discourse on the Origin of Inequality* (1754):

> And how shall man hope to see himself as nature has made him, across all the changes which the succession of place and time must have produced in his original constitution? How can he distinguish what is fundamental in his nature from the changes and additions which his circumstances and the advancements he has made have introduced to modify his primitive condition?

The answer would become apparent: by proposing a model of the natural man, with which to contrast modern man, and thus describe the ascent to civilization and humanity. And that model, once again, would be embedded in the indigenous inhabitants of the New World. Rousseau explains, for example, that "men in a state of nature are confined to what is physical in love" and thus are not subject to the fits of passion and jealousy so consumptive in modern society. And he identifies "the Caribbeans, who have as yet least of all deviated from the state of nature, being in fact the most peaceable of people in their amours."

In his essay on *The Social Contract* (1762), Rousseau describes the origin of social and political relations. "The first societies," he tells us, "governed themselves aristocratically. The heads of families took counsel together on public affairs. The young bowed without question to authority of experience. . . . The savages of North America govern themselves in this way even now, and their government is admirable."

Rousseau, of course, was romantically admiring the natural state, just as Hobbes was demonizing it. What they shared were two premises: (1) that there was a kernel of nature to be revealed beneath the layers of culture; and (2) the way to access it was through the most cultureless societies known, supposedly those of the Amerindians.

And thus the study of human nature became the study of the Indians. And through the nineteenth century, students of Indian languages would imagine they were studying something primitive and similar to the primordial or "natural" language of man; and students of Indian society would imagine they were studying the basic, and most rudimentary, manifestations of the social instinct in humans. To some extent this reinforced the idea that Indians should be assimilated—in other words, that layers of civilization could be imparted to them—and this affected federal policy toward the Indians.

From the standpoint of social science, however, the nineteenth-century Native American ethnology of Americans was invariably entangled with revelations of basic human nature.

That is, until the first studies of Australians.

Influenced by the renowned researches of Lewis Henry Morgan on the Iroquois, and more generally on the evolution of society, Lorimer Fison and A. L. Howitt published the first ethnographic monograph on the Kamileroi and Kurnai of Australia in 1880. Shortly thereafter, Baldwin Spencer and F. J. Gillen published their first monograph on the Australian Arunta.

In short measure, the Australian aborigines came to replace the Indians as the embodiment of "natural man"—human life at its simplest. Thus the French anthropologist Emile Durkheim would write his groundbreaking synthesis on *The Elementary Forms of the Religious Life* (1912) based on the knowledge that the most "elementary forms" of religion were those practiced by Australian aborigines—human religion at its most natural.

The years immediately after World War II saw an attempt by UNESCO to universalize humanity in the wake of the overzealous racialism of the Nazi era. Anthropologists had begun to conceptualize human history in terms of revolutions in subsistence—from hunting and gathering, to agriculture, to industry. And the group that emerged as the quintessential remaining hunter-gatherers were the !Kung San of the Kalahari Desert in southern Africa. From 1963–74, they were the subject of an intense interdisciplinary study organized through Harvard University. With the emergence of sociobiology at

Harvard, the biophilosophical problem of "natural man" once again came to the fore.

The !Kung San slipped neatly into the space formerly occupied by the Indians and Australians. A major conference on "Man the Hunter," organized in 1966, acknowledged the diversity of hunter-gatherers, but the organizers of the conference were also the organizers of the !Kung San study, and in short order, the !Kung became anthropology's paradigmatic hunter-gatherers. And by that fact, they became the representatives of "natural man"—particularly to the newly emerging field of human sociobiology.

The justification was that we had been hunter-gatherers for 99% of our existence as a species—food production and its social consequences had only been around for a few thousand years, and industrialism for a much shorter time. The economic system of hunting and gathering was what had shaped our gene pool.

On the other hand, we were fish for over half of our existence as vertebrates, but there doesn't seem to be a great deal relevant in the detailed study of fish biology if you're interested in humans. So the fact that the !Kung San shared certain broad economic and organizational features with Mesolithic Europeans may be irrelevant to the understanding of the behavior of Mesolithic Europeans, who had their own environment to adapt to, and their own traditions.

More than that, the !Kung San might not be valuable as "natural" people, either. Eric Wolf's *Europe and the People without History* (1982) argued persuasively that the mere fact that we don't know the histories of nonliterate, indigenous peoples doesn't mean that they have always lived exactly as they do today. Archaeological studies have found evidence of trade everywhere, which means that nobody was really completely "isolated," and ethnohistory has consistently found that even hunters and gatherers often have long and complex relationships with their neighbors.

By the 1980s, the !Kung San were the subjects of a fierce revisionist ethnohistorical battle. Sociobiologists continued to represent them as surrogates for "early man," living the edenic life we all evolved for. Ethnohistorians, on the other hand, found a far more interesting history of the !Kung San than the biologists had expected. It was a history of contact with pastoralists, European traders, colonialism, and economic exploitation: far from being pristine and standing outside of world history, and representing human nature minimally affected by culture, the !Kung San were very much products of that very world history. Their lifeways constituted revelations not about "people," but about *them.*

By 1992, Richard Lee, co-organizer of the !Kung San project, was seeing the endeavor in a very different light.

> In the preface to *Man the Hunter,* [Irven] DeVore and I wrote, "We cannot avoid the suspicion that many of [the contributors] were led to live and work among the hunters because of a feeling that the human condition was likely to be more clearly drawn here than among other kinds of societies" . . . I now believe this is wrong. The human condition is about poverty, injustice, exploitation, war, suffering. To seek the human condition one must go . . . to the barrios, shantytowns, and palatial mansions of Rio, Lima, and Mexico City, where massive inequalities of wealth and power have produced fabulous abundance for some and misery for most. When anthropologists look at hunter-gatherers they are seeking something else: a vision of human life and human possibilities without the pomp and glory, but also without the misery and inequity of state and class societies.

The !Kung San were no more primordial or natural than anybody else. The questions Rousseau had asked two hundred years earlier could not be answered by the hunter-gatherers of the Kalahari Desert.

In the shadow of the !Kung San fell the Yanomamo of Venezuela and Brazil, subjects of continuous study since the 1960s, when they became the subjects of a landmark genetic survey as a "primitive" tribe. They became mainstays of sociobiological generalization for their primordially aggressive ways and the discovery that the more aggressive men among them fathered more children. In fact, the generalization remained even after both assertions had effectively been refuted. To the extent that the Yanomamo really are aggressive, there's nothing primordial to it; they fight over caches of goods distributed by anthropologists and missionaries, not unlike the way people at the local shopping mall would fight over $100 bills raining down upon them. And the statistics simply don't bear out the assertion that aggressive men have more kids. In among a host of speculative accusations, Patrick Tierney's book *Darkness in El Dorado* (2000) effectively shows that the Yanomamo are anything but "natural"—they've been pawns for scientists, politicians, and missionaries, and their social history is complex, and often tragic—but as with the !Kung San, there's nothing primordial or natural about them. They're just exotic and remote.

And exploited and in poor health.

Fueled by the unquestioned assumption that the question was at least a valid one, the 1990s found a new surrogate for natural man. Rousseau's goal had been to "strip this being, thus constituted, of all the supernatural gifts he may have received, and all the artificial faculties he may have acquired only by a long process; [to] consider him, in a word, just as he must have come from the hands of nature. . . . I see him satisfying his hunger at the first brook; finding his bed at the foot of the tree which afforded him a repast; and with that, all his wants supplied."

That surrogate natural man, supplanting the !Kung San and the Yanomamo, would be the chimpanzee.

The chimpanzee had two things in its favor. First, misinterpretation of molecular data in the 1980s had crudely suggested that gorillas were not so closely related as humans and chimps were to one another. Thus, chimpanzees could be claimed as our sole closest living relatives, and the complications ensuing from considering the gorilla too, which is in many ways different from the chimpanzee, could be ignored. And second, the dehumanizing discourse that robbed the !Kung San and Yanomamo of their history and their cultural complexity could be applied to the chimpanzees less offensively. Chimpanzees were, after all, not human, not historical, and not cultural.

In *Demonic Males* (1996), primatologist Richard Wrangham takes up Rousseau's task. To Wrangham, the key feature that makes chimpanzees primordial people is their violent nature. Curiously, this violence was sufficiently subtle as to have been missed for the first few decades of chimpanzee research, in which conflict was noted, but not more than in other primate species. In the 1970s, at Gombe, however, the males of the Kasakela community killed off the Kahama community—for reasons unknown, but fundamentally and meaningfully similar, according to *Demonic Males,* to human doings at Gaugamela, Actium, Agincourt, Balaklava, Vicksburg, Ypres, Nagasaki, and Sharpeville.

And were not merely a projection of human cultural values onto another animal, like *Aesop's Fables* or Dogs Playing Poker.

And although he neglects to mention Thomas Hobbes, who started it all, Hobbes permeates this view of the chimpanzee as natural human: "[M]odern chimpanzees are not merely fellow time-travelers and evolutionary relatives, but surprisingly excellent models of our direct ancestors. . . . [C]himpanzee-like violence preceded and paved the way for human war, making modern humans the dazed survivors of a continuous, 5-million-year habit of lethal aggression" (p. 63). And so we come full circle. Humans again live naturally in a state of "Warre," but 350 years after Hobbes, that state is proved by recourse to evolutionary biology, and specifically by recourse to the behavior of the most natural man, the chimpanzee.

And yet, is there a biological evolutionary argument here? Even if we were to concede the (inaccurate) point that chimpanzees are genetically our sole closest relatives; the (contestable) point that the extinction of the Kahama community in the mid-1970s was a "normal" facet of chimpanzee behavior, although nothing like it had ever before been noted; and the (equally contestable) point that this implies that chimpanzees are naturally "violent," what might that imply for humans biologically?

As is obvious from the foregoing discussion, exceedingly little. It might indeed be a facet of human nature; or alternatively an element of specifically chimpanzee nature and never even present in a human ancestor; it might have been present in a human-chimp ancestor and lost in humans. You could study chimpanzee behavior until the Second Coming, and never be able to differentiate among these alternatives.

And that is a crucial point. The ability to distinguish between alternatives is testability, and it is a defining property of science. Without it, you're not talking science.

As a result, the discussion of human nature based on the properties of the chimpanzee lies in the realm not of scientific discourse but of metaphysics. It's equivalent to asking about Mozart's baseball talent—you cannot tell from the data available. A scientist might have an opinion about it, just as a scientist might have an opinion about whether Mike Tyson could have beaten Rocky Marciano, whether there is intelligent life in outer space, or how many angels can sit on the head of pin. But the basic inability to differentiate among the possibilities with real-world evidence makes this a nonscientific issue.

"Wars tend to be rooted in competition for status," *Demonic Males* is reduced to arguing. "We could well substitute for Sparta and Athens the names of two male chimpanzees in the same community, one rising in power, the other anxious to keep his higher status" (p. 192).

I suppose we could. But that would have no biological meaning. Who is competing for status—the generals? The grunts? The politicians back home? The sociopolitical entities they represent? And what is the biological connection between a chimpanzee and Greek city-state, anyway? A *polis* isn't a biological entity; it shares no common ancestry with a chimpanzee; it is an artifice of human social history.

The argument is simply an analogy. Maybe a chimpanzee is sort of like a Greek city-state. Maybe an aphid is like Microsoft. Maybe a kangaroo is like *Gone with the Wind*. Maybe a gopher is like a microwave oven. There may be arguments to be made in each case, but they are literary, not biological.

The most interesting feature of the argument for using the chimpanzee as a surrogate for "natural man" is the way in which the same animals are used to argue for very different qualities of human nature. Sociobiologist Frans de Waal argues from the bonobo, or "pygmy chimpanzee," that humans are naturally "good-natured." The fact that bonobos and common chimpanzees are *each other's* closest relatives should attest to the breadth of inferred natures possible in these animals and the speciousness of taking any one of them as natural for humans.

The answer is now framed in Darwinian terms, but the question is still the one asked by the pre-evolutionist Rousseau—How can we come to know natural man, our essence prior to the development of all these cultural complications? The quest for natural man has now become the quest for our precultural ancestor, who is taken to be "Us, But Just Without Culture." And that is supposedly revealed by the chimpanzee.

NATURAL SEX

The show was ABC's *Nightline*. Ted Koppel was off, and Forrest Sawyer was substituting. It was August 3, 1992, and the subject was the bonobo, in particular its extraordinary sex life. Extraordinary, that is, for an ape: female bonobos rub their genitalia together, stimulating one another, and they use sex in socially instrumental ways—and, more important, in nonreproductive ways. What might that tell us about human nature?

Professor Frans de Waal, a distinguished primatologist from Emory University, spoke quietly and candidly, in a mild Dutch accent, and explained how the meaning of what was being discussed in bonobos meant that "things that we still have today, such as pedophilia and homosexuality and a lot of other sexual orientations, that they are sort of leftovers of that earlier period, in which sex was more widely and publicly used in the society."

Now there's a thought that needs to be, as they say, "unpacked." Ignoring the gratuitous association of sexual preference and pedophilia, a sex crime, let's just focus on the last clause. The observations in the bonobos are first made relevant to humans by representing them as equivalent to their human manifestations. This in itself may be no more than a word game, for the equivalence of "homosexuality" in humans and chimpanzees might be like that of "slavery" in humans and ants, or "flight" in ducks, dragonflies, and airplane pilots—which have, genetically, developmentally, and evolutionarily, nothing to do with one another. Secondly, these bonobo and human attributes are both equated with an attribute of human ancestors, by definition unobservable. In this quite static evolutionary view, the attribute thus becomes an unchanged survival from ancient times, a chunk of human nature made manifest.

That is simply pre-evolutionary philosophical thought clothed in evolutionary terms.

We see this reasoning likewise in *The Hunting Apes* by sociobiologist Craig Stanford. "[C]himpanzees, ancestral hominids, and modern foraging people such as the !Kung or Aché—provide a frame of reference of our evolutionary history and therefore the roots of human behavior," Stanford says. Thus we are presented with another species, our own ancestors, and some modern humans as all being sufficiently equivalent as to be effectively interchangeable. To study one is to study the others; to study chimps is to access human nature.

A DUCK WITH LIPS?

They used to teach a course in the biology department at Yale in which the professor would raise the question of how to study human nature without the confusing overlay of culture. He assured the class that this was a modern scientific question, and not simply a rhetorical device from political philosophy of the seventeenth and eighteenth centuries.

We discussed it over lunch one day, and he patiently explained it to me as well, that to study humans scientifically, we had to examine them *as if they had no culture.* Culture, he said, only confuses things.

He was an ornithologist.

I finally replied, "Asking what humans would be like without culture is like asking what a duck would be like if it had lips instead of a bill. It wouldn't be a duck; it would be something else."

I don't know if he ever got it. We never had lunch again.

On the other hand, that course is no longer offered.

Rousseau and Hobbes were wrong, as was the ornithologist. There is no human nature outside of culture. A human is not like an artichoke, in which the prickly leaves of culture can be peeled away, leaving only the exposed tender heart of "natural man."

To define human nature is to define as well human un-nature. If carnivory shaped our evolution, as one sociobiologist has written, then being a vegetarian is unnatural. And yet people survive, and thrive, and quite happily, as vegetarians. If polygyny is human nature, then monogamous people are unnatural—and yet many are happy that way.

Human nature is merely the range of things humans are capable of. Some are noble, some are destructive, most are merely variants. But if humans can be found to be both peaceful and aggressive, say, then it makes little sense to speak of one or the other as being human nature—and it's a ridiculously trivial observation to say that *both* are human nature. That is because the question is framed archaically.

Nature and culture act as a synergy. If the human is like a cake, culture is like the eggs, not like the icing—it is an inseparable part, not a superficial glaze. Whatever humans do, or look like, is a product of both. Consider:

What is the shape of the human head? Is it globular or elongated? You'll find those extremes, and everything in between, atop the shoulders of normal people. We know that head shape can be modified culturally—the Incas bound the heads of their children to give them high skulls, and the Flathead Indians were renowned for skull modifications in the opposite direction. The use of cradleboards affects the shape of the back of the skull. But culture has far more subtle effects, as Franz Boas showed early in the twentieth century. Merely growing up in America, instead of in eastern Europe or Sicily, affects the form of the skull. We don't know how, or what specifically are the critical variables. But as we noted earlier, this simply means that the human body develops and grows responsively to the conditions of life. You cannot develop a human skull outside of its environmental boundary conditions. And in humans, those conditions are cultural.

Is it human nature to be polygynous? In primate species characterized by a male mating with several females, the males are generally substantially larger than the females. Human males are on average 20–25% larger than females (but this can obviously be exaggerated in cultures where boys are encouraged to be big and strong and girls are encouraged to throw up lunch). Perhaps, then, this means we are naturally somewhat polygynous, as sociobiologist Jared Diamond has suggested in *The Third Chimpanzee* (1992)? On the other hand, males in polygynous primate species have canine teeth substantially larger than those of females; in monogamous primates (such as the gibbon), males and females are not only the same size, but also have canine teeth the same size. What's the situation in humans? Surprise! Males and females have canine teeth of the same size. (Diamond neglects that counterevidence in presenting his case for human nature.) What's even more confusing is that humans have differences in body composition, with women averaging more fat and men more muscle; and differences in facial and body hair—patterns of male-female differences without comparability in other species, for they are not seen in our close primate relatives. So one bit of data suggests polygyny, another suggests monogamy, and a third suggests that no comparative inference about human nature, narrowly construed in a single noun like "polygyny" or a single adjective like "demonic," is possible. Clearly, diverse sociosexual forms coexist in the human species; human nature appears to be everything and everywhere.

Is it natural to menstruate? In half our species, once a month a follicle is released into the fallopian tubes, is not fertilized, and is discarded with the endometrial lining of the uterus. What could be more natural? And yet, when we compare our culture to others, we find that women in modern industrialized society become fertile at an earlier age, have fewer children in their lifetimes, space them more closely together, don't take full advantage of the hormonal suppression of ovulation that accompanies intensive breast-feeding, and reach menopause at a later age. When you do the math, you discover that women in our modern society have five to ten times as many menstrual cycles during their lives than women in rural, nonindustrialized societies, and *far more menstrual cycles than women at any other time, in any other place, have ever had.* In other words, cultural factors such as economics, diet, and life expectations combine to determine the number of menstrual cycles a woman will experience in her life.

What we get from all this is that the quest for human nature is itself an illusion, the result of an antiquated way of thinking about humans scientifically—with culture as a veneer overlying nature. Every human being that has ever lived has been born into a culture. Culture preceded our species—our ancestor *Homo erectus* certainly was a cultural species. In other words, culture is part of our nature; you can't strip it away, scrape it off, pretend it doesn't exist, or look for it in simpler forms. It is ubiquitous in human life; it is *programmed into* human life.

You can't derive human nature from the Indians, the Australians, the !Kung San, or the Yanomamo. Each is merely a variation on the human theme—a theme as broad as the diversity of human societies, languages, and activities over the span of the globe and the millennia (and which, as it is reduced through the penetration of genocide, colonialism, the media, and the market, reinforces the illusion of uniformity). But none of them reveals human nature any more clearly than a Harvard faculty meeting does. As Thomas Hobbes realized when he invented this line of reasoning, it's simply a series of political inferences invoking the authority of science.

More than that, you can't get at human nature from chimpanzees. They're not human.

Eight

HUMAN RIGHTS . . . FOR APES?

WHEN WE TALK ABOUT WHETHER antisocial behaviors are innate or not, we are making anthropological pronouncements and invoking what looks like genetics in support of a social and political philosophy. In this case, the idea is that the poor and uneducated deserve what little they have, for nature (rather than the avarice and malice of others) has put them in the social position they are in. This can be easily recognized as the central argument of *The Bell Curve,* for example. But much of Western social history has involved the development of views antithetical to that one, social ideologies that are somewhat more empathetic.

We lack tests to gauge the individual potentials that people are born with, and cannot make scientific statements about them. In the absence of a scientific approach to human potential, we are left with two poles of social action: we can try to identify and cultivate diverse talents as widely as possible (by investing our resources in education, day care, and children's enrichment programs); or we can condemn large groups of people on the basis of their lack of performance, denying the right of the subjects to be judged as individuals, and denying as well the possibility of society at all benefiting from the talents with which they are endowed.

Neither alternative is particularly scientific; but the first is at least humanitarian.

There are other venues, however, by which social and political philosophies invoke molecular anthropological data in support of their positions. As we saw in chapter 7, sociobiologists make plenty of hay from the genetic similarity of humans and apes. One way in which they extend this genetic similarity is to redefine terms and concepts generally applied to humans so that they can now apply to the apes as well.

CULTURE

In the previous chapter we saw how the 98% genetic similarity is used to construct chimpanzees as the "simplest humans"—a role no longer filled by humans of any sort. Now the argument can be extended to argue that the chimpanzees actually do possess in some rudimentary state the quintessentially human attribute—culture. If they have that, which is supposed to be the thing that distinguishes humans from other creatures, then they could really be considered simple people, in a way.

Now, culture is a very tricky concept. It was invented by German anthropologists in the nineteenth century and introduced independently to the American anthropological community by the Englishman Edward Burnett Tylor and the German American Franz Boas. Tylor used the term in 1871 as synonymous with "civilization" and to mean the things that aren't inborn and of which all peoples partake to a greater or lesser extent: "the capabilities and habits acquired by man as a member of society."

Boas, for his part, democratized culture and broadened it beyond "civilization," but never got around to trying to define it until very late in life. When he finally did, it came out a bit less felicitously than Tylor's definition: "the totality of the mental and physical reactions and activities that characterize the behavior of the individuals composing a social group . . . in relation to their natural environment, to other groups, to members of the group itself, and of each individual to himself. It also includes the product of these activities and their role in the life of the groups."

Boas significantly went on to say that "[t]he mere enumeration of these various aspects of life, however, does not constitute culture." Culture was something more than just the sum of its parts. After all, "[t]he activities enumerated here are not by any means the sole property of man, for the life of animals is also regulated by their relations to nature, to other animals, and by the interrelation of the individuals composing the same species or social group."

In practice, what makes culture more than merely the enumeration of the sum of its parts is *language,* the fundamentally symbolic activity that permeates and suffuses all forms of human activity.

And that's the way anthropologists generally used the term in the twentieth century.

That has been challenged recently in two ways: arguing that chimpanzees make and use tools, and have local traditions, and therefore culture; and arguing that chimpanzees have language.

One thing chimpanzees do not have is speech; but speech isn't language. Mynah birds speak, but they don't know what they are saying. Speech can be an essentially mindless act of sound production; but language is a mental faculty. Mute people are still human, but mynah birds aren't. Chimpanzees, for all their genetic similarity to us, cannot even begin to approximate the range of sound that a person can; indeed, that a mynah bird can.

Experiments with sign language and with computers, beginning in the 1960s, were designed to see just what chimpanzees are mentally capable of, along a linguistic path. Although they can't speak, perhaps given a different medium than vocal, they might be shown actually to approximate human faculties.

For all the interest generated by the sign-language experiments with apes, three things are clear. First, they do have the capacity to manipulate a symbolic system given to them by humans, and to communicate with it. Second, unfortunately, they have nothing to say. And third, they do not use any such system in the wild.

This is not at all to disparage the value of that research. Given the genetic and evolutionary proximity of the apes to us, it would be somewhat surprising if they did not show any similarity to human cognitive skills. But the studies were disappointing, for the apes never said anything like, "Would you like to know about our social system? Well, the big guy, whom you call the alpha male, stomps around and the offspring, who may or may not be his, get out of his way, unless their mother is around."

Instead, they said, "Come, tickle now. Gimme sweet."

It's also pretty clear that apes are very smart and can indeed make and use tools—cruddy tools to be sure, but tools by anyone's definition—and that they can communicate coarsely in a human venue.

But that's not what anthropologists mean by culture.

Nor is the recognition—the subject of an editorial in the *New York Times!*—that chimpanzees have local traditions, "the ability to pass on the results of observation and learning." No sane modern scholar would deny that animals observe and learn. Culture is learned, but it doesn't follow that everything learned is cultural. The fact that humans have hair does not imply that all things with hair are human. That's from Philosophy of Logic 1A. But it comprises the core of this odd argument: since culture is learned behavior, and chimps learn behavior, it follows that chimps have culture—in spite of what anthropologists have thought about it for decades.

And in spite of Boas's own words decades ago.

What this revelation now shows, according to the *Times*'s editorial, is that chimps "are more like us than we care to believe, and we are more like them than we like to let on."

Well, of course, that's true whether or not they are cultural. But what's the point of asserting that they are? Especially if it means redefining "culture" so as to make the demonstration that chimpanzees have it trivial?

Three separate questions must be answered to make sense of all this:

What do humans have?

What do chimps have?

What's their relationship?

What humans have is a zoologically unique way of communicating, which imparts arbitrary meaning to sounds, to combinations of sounds, to the order of those combinations of sounds, and to the associated social contexts of utterances. Our communication involves phonology—the sounds, and especially "decisions" about them, such as "the rolled r and the trilled r, which are different in Spanish, shall be variants of the same sound in English"; semantics (words or parts of words); syntax or grammar; and sociolinguistics (for example, the knowledge that when Steve Martin said, "Well, excuse me!" in the early days of *Saturday Night Live,* it was not intended to be a polite apology, even though it appears to be one on paper).

This system, involving multiple levels of arbitrariness, is unique to humans. It is so well developed in all human societies—symbolism to the fourth power!—that it is difficult even to imagine how it might have evolved, although it obviously did.

Its power permits us to discuss things that didn't happen, that might happen, that will happen; things to make us happy or sad; to induce sexual arousal or exhortation to combat; to praise or berate one another; to persuade, to lie. A narrowly utilitarian approach emphasizes the capabilities of transferring information from one person to another, or to many people concurrently, but neglects the many other kinds of things humans have to say to one another—and probably have always had to say to one another.

And that chimpanzees don't, and haven't.

Which is not to put chimps down. We in fact know rather little about how they communicate with one another; we've given names to their sounds, like "grunt" and "pant-hoot," which occur in specific contexts. But what they do have in the way of a communication system is not well known; and what they have in the way of cognitive apparatus—the ability to use a human system of communication— appears to be present, but about as useful to a chimpanzee as the echolocation of bats would be. Language is just not a chimpanzee thing. There is in fact very little overlap between chimpanzee and human communication.

The news that they have "culture" depends crucially, then, on your understanding of what is meant by the term. In this case, the scholars who published the paper in *Nature* that got the attention of the *New York Times* meant something very specific. To them, culture is a set of learned behaviors or traditions, which vary from group to group.

And they showed that chimps have that.

Not that it would have come as much of a surprise to Darwin or Boas. I'd venture to guess that populations of many animals have learned behaviors that are different from those of other populations.

So why all the fuss?

Humanizing African apes—just like *de*humanizing African *people*—has political weight. It might serve to make people aware of the abuses chimpanzees sometimes suffer in captivity; it might arouse sympathies for the animal rights movement; or it might just serve as yet another rhetorical weapon against the fundamentalist Christians still trying to subvert science education in America.

THE GREAT APE PROJECT

A few years ago I was invited to Manchester, England, to participate in a TV film that would debate the merits of a movement to generate interest in giving human rights to the great apes, The Great Ape Project.

The Great Ape Project grew out of an apparent convergence of interests between primate conservationists and animal-rights activists. On the one hand, there is a worldwide problem that has not yet been effectively addressed. Apes are threatened in the wild by the economic development of the human societies around them. Part of this threat is their exportation, legal and illegal. Apes are often objectified by callous and cynical entrepreneurs, who neither regard them nor treat them as the sentient, emotionally complex creatures they are. They are generally disposed of when they lose their cuteness, usually less than one-fifth of the way through their lives. The lucky ones can live out their lives in the care of an enlightened or sympathetic zoo or primate research facility with sensitive caretakers and handlers. Most, of course, are not lucky.

On the other hand, they strike the most self-reflexive chord in us when we look at them. They are our closest relatives; they sort of look and act like us, and they certainly have nervous systems, reactions, and behaviors like our own. So to the extent that animals are "persons" deserving of rights, these are the animals most deserving of all.

The Great Ape Project proposes that the great apes are entitled to life, liberty, and not to be tortured—in other words, to the rights of humans, with whom they form "a community of equals." Fair enough; after all, who these days would stand up in favor of death, enslavement, and torture? In the early 1990s, a book called *The Great Ape Project* adduced no fewer than thirty essays in support of the idea of human rights for the great apes. But in its way stand two simple facts, which I tried to articulate for the British television audience.

First, the apes aren't human.

And, second, we can't even guarantee human rights to humans.

Apes, being endangered, are no longer imported from the wild for scientific research in America. We can readily acknowledge that they deserve humane treatment in captivity. But *human* treatment? To whom are they equal, anyway, and on what basis can we say so?

The fundamental basis, once again, was asserted to be their genetic similarity to humans. This point was authoritatively asserted by the biologists Jared Diamond and Richard Dawkins in their contributions and echoed by several others. It was also the very first point addressed in the filming of the television show. The molecular factoid thus became the basis for a push for social legislation and moral reform.

But we need to start out by remembering that we just do not accord rights in modern society on the basis of genetic distance. The English do not accord greater rights to Iraqis than to Japanese. The category "human" is neither fuzzy nor defined by genetic closeness. In principle the category should be defined (as any animal species should be) in terms of reproductive compatibility and ecological niche—not in terms of how genetically similar or dissimilar two individuals may be.

More to the point, the tenets of liberal democracy since the Enlightenment hold that citizens should be entitled to equal rights. Of course, as philosopher Hannah Arendt noted, the Enlightenment had it backward in arguing that "all people are created equal," which entitled them to citizenship under the law. Actually, however, citizenship is neither an act of God, nor a fact of biology, but an endowment by the state. It is specifically the conferral of citizenship that makes people equal. The state decides who its citizens are and makes them equal. And to the extent that since the end of World War II we have sought to broaden the concept into one of universal rights, that is predicated on the idea of universal citizenship.

To suggest that genetic distance should now dictate the allocation of rights harks back to some of the more odious social philosophies of this century. Certainly human rights agencies are not concerned with how similar people are to one another; they are concerned with the maltreatment of people. The entry of genetics would seem to be a perverse change of moral direction.

Moral direction is notoriously fickle, though. My initial thoughts about the push to give chimpanzees human rights harked back to the 1950 movie (and recent musical) *Sunset Boulevard*. The narrator, a young screenwriter, played by William Holden, is visiting the home of the reclusive screen legend Norma Desmond (Gloria Swanson). His first glimpse of her is as she cradles and mourns her lately expired pet chimpanzee. Watching "the last rites for that hairy old chimp, performed with the utmost seriousness, as if she were laying to rest an only child," the Holden character wonders: "Was her life really as empty as that?"

Obviously, it was. We learn of her emotional isolation and her inhumanity through the course of the movie. And foreshadowing the dementia ("All right, Mr. DeMille, I'm ready for my close-up," she coos famously as she is led out) is the fact that she has lost track of her own kind; she doesn't know what she is. She treats animals like humans and vice versa; she has no moral compass.

Fickle indeed are the moral winds. When *Sunset Boulevard* was released in 1950, Norma Desmond's confusion of ape and person was an obvious marker of her loss of her moral, and impending loss of mental, bearings. Today, the same confusion is worn as a badge of morality.

Fickle, indeed!

Apes deserve protection, even rights, but not human rights. Confusing humans and nonhumans has never advanced anyone's lot. And certainly genetics is not going to clarify it much.

By the genetic criterion, the cutoff (at orangutans) is quite arbitrary anyway. Indeed the ape-rights movement is aimed specifically at the *great* apes (chimpanzee, gorilla, orangutan), and leaves the graceful and endangered *lesser* apes, the gibbons of southeast Asia, to their own devices.

But this is where the alliance between the animal-rights activists and the ape-conservation activists beaks down. The former group sees the ape-rights movement as a Trojan horse, a foot in the door toward the liberation of all animals; apes just happen to be among the cutest and most resonant, and thereby the ones for which they can drum up support most easily. Some of these supporters of the ape-rights movement don't even seem to know much about apes; one essay in their flagship document actually begins, "Chimpanzees make love rather like humans do. . . ."

Of course, it is not clear that chimpanzees "make love" at all. But more important, the average copulatory bout in chimpanzees lasts between ten and fifteen seconds and involves a female whose genitalia are swollen and purple (a prominent visual display of fertility, unique among apes to the chimpanzees), and often a succession of males who are otherwise generally not very interested in sex; and no tactile exploration or erogeny at all. In ten to fifteen seconds, after all, you can't expect much. Doesn't sound the least bit like love-making in humans, at least the way most of us been fortunate enough to experience it, and it quite possibly says more about the author than it does about chimpanzees.

But for the extreme animal-rights activists, genetic distance doesn't much matter to begin with; it is simply a rhetorical device. If all animals (or at least, mammals; or at least, cute mammals) are equivalent, then the apes hold no special place for them, and the evolutionary relationships between us and them are entirely irrelevant. The conjunction of animal-rights activists and sincere primate conservationists is thus a marriage of convenience. If the justification for ape rights is phylogenetic proximity, then the justification for "animal rights" is creationist, for it ignores evolutionary relationships and implies that all animals are equidistant from humans.

As the editors of *The Great Ape Project* write, the most important contribution of the political push "will be its symbolic value as a concrete representation of the first breach in the species barrier." This sounds like an exploitation of the living apes and of their difficult status as callous as any circus act is.

ARE APES MERELY DISABLED PEOPLE?

One significant argument advanced on behalf of human rights for the apes involves their cognitive performance. Since their brains are closely related to our brains, it should come as no surprise that the apes can approach humans in their cognitive functions. To the extent that those functions can be linearized for comparison, apes perform mentally on a par with some mentally disabled children.

But that is a very misleading comparison, for two major reasons. The fact that tests can be linearized does not mean that the relationship of apes to us is itself linear. Apes are not *sub*human, they are *non*human. Thus, although the largest gorilla brain is similar in size to the smallest nonpathological human brain, there is no evidence that the gorilla brain ever produces anything other than normal gorilla thoughts or that the human brain produces anything other than normal human thoughts.

But the comparison is even more invidious, for it often reinvents the linear developmental hierarchy of Victorian science. A century ago, evolution was commonly seen as a single-dimensional process, with healthy adult European Christian men at a pinnacle, and non-Europeans, non-Christians, non-men, non-adults, and non-healthy people somewhere below them. In essence, the other categories of people comprised humans who were not fully developed; to the extent that they were at least biologically human, they were nevertheless incomplete.

Over the course of the nineteenth and twentieth centuries, that view was eroded by the elevation of all those other people. If healthy adult European Christian men were at the pinnacle of something, it was a social hierarchy that placed them and kept them there by exploiting and subjugating other people. The problem was the agency by which inequality came to be regarded as acceptable, and even natural—not the qualities of the "incomplete" people of the world. This was not a natural hierarchy, expressing the superior qualities of those people at the top; that view was a classic case of blaming the victim.

Curiously, we see that hierarchy rediscovered in the Great Ape Project. A British zoologist was asked on our television show, as a supporter of human rights for apes, whether he would allow his daughter to marry one. He replied, "I don't think she would want to marry one. . . . Would you let your daughter marry, say, a mentally retarded individual?"

Other members of the discussion gasped. Did he really think that there was some kind of equivalence between a mentally handicapped person and a chimpanzee? Is marrying outside your species really equivalent to marrying outside your IQ cohort?

It was a chilling equivalence, made very casually.

Not only are apes not at all like mentally handicapped humans, but more important, *mentally handicapped humans are not like apes.* Nor are blacks, Jews, hunter-gatherers, or New York cabbies. All people are perfectly human, and all apes are not at all human. That's the simple biology.

And yet there is a seductive ring to the comparison. As a legal advocate for the movement said on the television show:

> We give human rights to children, to the aged and to the mentally infirm, to the autistic, to the deaf and to the dumb. . . . [T]he facts we recognize in common humanity compel us to recognize the humanity in the great apes; . . . they can reason and communicate at least as well as some of the children and disabled humans to whom we accord human rights.

"Humanity" here seems to be something a bit unfamiliar, something other than merely "the state of being human," which the apes clearly are not and disabled people clearly are. It seems to be no longer a condition of birth, a simple fact of nature or genetics, but rather a state one can aspire to, and may earn through industry and mental vigor. If it can be earned, of course, it can also be lost; they are two sides of the same coin. And so the key question is: What exactly is the implied connection between the ability to reason and communicate on the one hand, and humanity and human rights on the other? Why should the mentality of apes have any bearing on their humanness (or lack thereof) or their rights (or lack thereof)? If you lose the ability to reason and communicate, do you thereby forfeit your humanity and rights?

This is a scary moral place for apes and people to be. Humans have human rights by virtue of having been born human and to the extent that being a human is a precondition of citizenship. Human rights should neither be forfeitable nor accessible by nonhumans. That is not to say that other beings should have no rights; it is merely to say that the phrase "human rights" has no meaning if it does not apply to all humans and only to humans. Singling out particular classes of people in order to show how similar they are to apes is a troubling scientific strategy, not least of all when the humans rhetorically invoked are the very ones whose rights are most conspicuously in jeopardy.

The closest example I can think of is the fateful decision made in the nineteenth century by the great proponent of human evolution Thomas Huxley. Knowing how intensely political the subject was, Huxley let out all the stops in his public defense of evolution, particularly of human evolution. Unfortunately, Huxley was faced with an embarrassing lack of data from the fossil record: just where one would expect to see a transition from ape to human, there were no such fossils.

Not to worry, argued Huxley. Evolution is well supported in spite of the absence of fossil data: the transition from ape to human is documented among living peoples. The nonwhite peoples of the earth, argued Huxley, are the forms of life that connect the races of Europe to the ape as surely as the duck-billed platypus connects mammals to the lower vertebrates. The argument was familiar to creationists, because it had been invoked in pre-evolutionary writings; associating non-European peoples with lower forms of life transcended the argument of evolution versus creationism. But Huxley gave the association a new scientific context.

And so the non-European peoples were sacrificed on the altar of evolution in order to make a rhetorical point against the creationists. If we learn something from that episode, it must certainly be that dehumanizing people is not an admirable rhetorical strategy for a scientific debate.

The problem, as I see it, is that the great ape activists have tried to make this a "human rights" issue, to which they assume all humans would be empathetic. That, however, makes "human" the crucially contested concept, which it shouldn't be. The concept should be "rights."

Apes should be conserved and treated with compassion, but to blur the line between them and us is an unscientific rhetorical device. And thereby to call the human status of the most marginalized people in our society into question is morally problematic (in addition to being zoologically ridiculous).

We also acknowledge that alongside the concept of rights comes the attendant concept of responsibilities. You have a responsibility to behave in a civil manner, and can forfeit certain rights by the commission of criminal acts, for example. Of course, a chimpanzee cannot grasp the difference between civil and uncivil behavior, or between legal and illegal activities. A human would be institutionalized in such a state, but that is precisely what the ape-rights activists oppose for apes. How, then, do you treat a chimpanzee who steals or murders?

Or do you just accord them human rights without attendant responsibilities, in essence giving them greater license than humans?

Here the ape-rights activists are in a bit of a bind, for they try to represent chimpanzees as Jane Goodall did decades ago, as "nice" (and Goodall is a prominent activist for chimpanzees). But the latest work on chimpanzees shows them to be frequently mean and brutal, occasionally carnivorous, and always dangerous. Not quite the stuff of a sympathetic portrayal of human rights.

Perhaps the ugliest undertone in the discourse of "human rights for apes," however, is the combination of naive colonialism and cynical misanthropy. Ape rights is not a political movement in Central Africa or Indonesia, where *human* rights are sufficiently precarious. The ape rights movement is principally Euro-American.

When we hear talk of "human rights for apes" because they constitute a "community of equals," we are hearing a plea for the consideration of the interests of apes and humans to be given some kind of equal footing. Now don't get me wrong; nobody loves a cute baby chimpanzee more than I do, but when we think about ape rights, we have to remember that the ordinary New Yorker's or Londoner's economic interests simply do not come into conflict with those of a chimpanzee. The average Tanzanian's economic interests, on the other hand, may well.

The basic problem is the presumption that comes with a couch-potato argument for rights to Tanzanian chimpanzees, Rwandan gorillas, or Indonesian orangutans. Could you really look a group of Rwandans in the eye, with the horrors and brutalities on massive scales that they have had to endure, and tell them that a gorilla has got the same rights as them? Personally, I couldn't, but there are people who seem to think they could.

Gorillas are sentient and emotionally complex, but of course, so are humans. The people who would be most directly affected by such a declaration are those whose ordinary lives have been far crueler than those making the declaration. The welfare of gorillas is important, but it is not a human rights issue and cannot be allowed to diffuse the attention we pay to human rights violations.

THE BANGKOK SIX

In Bangkok airport in 1989, a scheme to smuggle six baby orangutans from Indonesia to (presumably) China was accidentally thwarted when the cries of the apes, in wooden crates in transit, alerted officials, after they had been abandoned for several days. One of the babies died a few weeks later, one a few months later, and a third a few years later. An appalling tragedy? You bet.

Could it be remedied by making the home country of orangutans, Indonesia, recognize human rights for orangutans? Maybe, but think about it. Indonesia is one of the worst places in the world for enforcing human rights for *humans*. On the island of East Timor, hundreds of thousands of people have been the victims of genocide in recent years. Indonesian attitudes toward orangutans are also tragic, but the simple fact is that state-mandated policies toward the apes simply do not constitute the biggest problem of rights there.

One occasionally encounters a curious response to this argument. As a noted British zoologist once told me superciliously, "Think percentages, not numbers." In other words, the Earth can spare a few hundred thousand East Asian people, because there are lots of us, but there are only a few orangutans, so each ape life is that much more valuable.

It is an odious comparison, weighing the value of human life against the value of orangutan life. But that's ultimately what this is about. A British professor thinks there are too many Asians and not enough orangutans.

Maybe, on the other hand, there are too many British professors.

This patently misanthropic ethic undermines any claim that ape-rights advocates might assert to the moral high ground. The value we place on a human life cannot be a function of how many others there are in that particular category. If you really believe there are too many people and not enough orangutans, and you value their lives accordingly, you are obliged to decide *which* people are superfluous.

Is your own life superfluous? Or just the lives of little dark people somewhere else whom you don't know?

That's often the way it is, of course. It is easy to condemn the poor of other nations, or even of our own, for being prolific, all the while knowing smugly that the world will always be big enough for one more Harvard man. That is in fact the fallacy promoted by Thomas Malthus at the end of the eighteenth century. He projected too many people and not enough resources and concluded we needed to take a heartless approach to the poor because *they* were the problem. The social forces that created the poor and kept them poor were invisible to him. Nowadays we know that education, upward economic mobility, and economic modernization all work to drive reproductive rates down; giving women choices in their lives invariably has them bear, on the average, fewer children.

It is far too easy to assert that there's too many of those other people; *they* can tolerate some losses. Not only is that callous, it is the very antithesis of morality in our civilization, and has been (at least normatively) for a very long time. Recall by way of mythic example the Pharaoh in the first chapter of Exodus, who decided there were too many Hebrews, and acted accordingly. A concern for animal welfare must come out of a concern for *human* welfare. It must emerge from a concern for human rights, not supplant it. For once we begin to devalue human lives, we lose a standard by which to value any other kind of lives. And it just doesn't work the other way around.

It is important—*very important*—to protect and defend the lives of nonhuman primates. But our concern for them can't come at the expense of our concern for human misery and make us numb to it. Baby orangutans in boxes should make you sad, but human misery should make you sadder. An abused chimpanzee may be genetically 98% human, but an abused person is 100% human.

The people who maintained overtly that most human life was cheap were the ones who *lost* World War II.

They were also the ones who thought that genetic distances constituted a sound basis on which to accord human rights.

READING THE MINDS OF APES

The most resonant argument for human rights for apes is based on their mental capacities, which are notoriously difficult to infer. Many of the data, and especially the emotionally compelling data, are anecdotal or based on a combination of projection and dubious inferences.

Just look at them! Just be with them—and you will experience the empathy that links your mind to theirs!

And yet, where clever, controlled experimentation has been possible, it has tended strongly to show that in specific ways, ape minds work quite differently from human minds. Which is, of course, what we would expect, given that they are different species.

Thus, argues Daniel Povinelli, you can watch chimps and imagine you know what they're thinking, but what you think they're thinking may not be a reliable guide at all to what they're actually thinking. Is a chimpanzee able to put itself in another's place, as it casually seems to human observers accustomed to imagining themselves in someone else's place, or is a chimpanzee's behavior guided by a less complex, less human mental program? Povinelli finds that whenever he can devise an experiment to distinguish between whether a chimpanzee is having a high-level thought or a low-level thought, the chimps invariably are having the simple, unimpressive thought—if thought is even the right word.

The data invoked for ape rights on the basis of their minds and behavior are invariably the fluffiest and least scientifically compelling data; where data are collected most rigorously, they point to the mental differences between us. Having three times as much brain does, it seems, make a difference.

None of which is to say that chimpanzees don't merit protection in captivity and preservation in the wild—but simply that it has nothing to do with human rights.

TAKING IT TO THE STREETS

In early 1999, the legislature of New Zealand debated the first ape-rights bill. There are no apes indigenous to New Zealand, and no research was being performed on apes in New Zealand. In fact, there were at the time twenty-eight chimpanzees and six orangutans in the entire country, and no one was suggesting that they were being in any way mistreated, so the proposed legislation was almost entirely symbolic. Appropriately, its major scientific proponent was a theoretical biologist.

Although largely ignored in the United States, the New Zealand debate was watched carefully in England, particularly by the scientific community. The weekly science magazine *New Scientist* weighed in strongly in their issue of February 13, reminding the reader that folk genetics lies just behind the legislation:

Unfortunately, it has become fashionable to stress that chimpanzees and humans must have staggeringly similar psychologies because they share 98.4 per cent of their DNA. But this misses the point: genomes are not cake recipes. . . . A creature that shares 98.4 per cent of its DNA with humans is not 98.4 per cent human, any more than a fish that shares, say, 40 per cent of its DNA with us is 40 per cent human.

And of the notion of human rights for the apes generally, *New Scientist* had only sarcasm:

[I]f animals have rights which protect them against humans, it is only logical that they should have rights that protect them from each other. If a chimp kills another chimp in the wild, or a human, do we really want to hire a fleet of lawyers? And if we extended honorary personhood to all animals, would the gazelle be entitled to rights against the lion?

The point, quite properly, is that such a movement invokes science for an entirely unscientific end. The means are rhetorical; to the extent that it sounds as if the end is supported by science, or by genetics, it merely makes the scientists look foolish or thoughtless.

Nature ran a small article in its issue of June 17, 1999, titled "NZ Bid to Win Rights for Apes Fails in Parliament." The idea of human rights for the apes died in committee. The chimpophiles had been thwarted. "This probably spells doom for their efforts to convince the United Nations to make a Declaration of the Rights of Great Apes, based on the genetic proximity of apes to humans."

Amen.

Nine

IF APE GENETICS IS POLITICAL, imagine how political *human* genetics gets. The big lie all too frequently taught in university genetics classes is that human genetics is, or at least ought to be, just like mouse or fly genetics.

It isn't.

It can't be. Mouse genetics doesn't have to deal with abortion, genetic screening, racism, discrimination, the loss of health insurance, and the rights of oppressed and indigenous mice.

Everyone knows about the Human Genome Project, a medical genetics program begun in 1989 for sequencing all the DNA in a typical human cell and identifying all its genes, and incidentally keeping every molecular geneticist in the world gainfully employed well into the millennium.

When you think about it, however, the Human Genome Project has a minor conceptual flaw at its center. Medical geneticists often conceive of genetic diseases in medical terms, in which people are either normal or ill—with a "normal" gene as the referent and its pathological variants as contrasts. But that's not necessarily the right way to see the human genome. After all (as noted in chapter 5), noses are all genetically determined, all different, and, barring accidents, all normal. Thus it is not appropriate to think of noses in the same way we think of diseases: a single condition of normalcy and a spectrum of deviations from it.

Similarly, the genes for blood groups, the longest-studied genetic system, are arrayed quite differently from the gene that governs cystic fibrosis. Rather than a normal gene and some rare disease-causing variants, all populations of humans have all the blood group alleles (A, B, and O) in differing (but high) proportions. This being the case, there is no "normal" blood group gene; type O, type A, and type B are all normal. And this seems to be the general pattern of human genetic variation, not the disease model.

Since people are genetically diverse, and most are nevertheless normal, there can consequently be no "*the* human genome" to sequence. Geneticist Bruce Walsh of the University of Arizona and I pointed this out in a sarcastic letter to *Nature* in 1986. There are many human genomes, all different from one another and all more or less normal. Genetic diseases are the exception, not the rule, when it comes to variation in our species. And if we are going to sequence one human genome, by implication, the "normal" one, perhaps—Walsh and I argued tongue-in-cheek—the Human Genome Project should exhume Charles Darwin and sequence *his* DNA. At least that might remind them of the flawed premise guiding their work.

But sarcastic letters will only get you so far. At stake might be real money, real prestige, and real jobs. Thus, population geneticists, hoping for a piece of the action, proposed a Human Genome *Diversity* Project (HGDP) in 1991. Led by Stanford's Luca Cavalli-Sforza, they proposed a major, centralized effort at establishing a genetic research museum of the human species. By establishing cell lines derived from the various peoples of the world, it would be possible to maintain a resource in Palo Alto or New Haven covering genetic diversity in the human species for study by geneticists anywhere, anytime.

Cavalli-Sforza, a charismatic and respected researcher in human population genetics, had been interested for decades in the biogenetic relationships among populations. For instance, given their genes, are the English more closely related to the Irish or to the French? Cavalli-Sforza had been developing algorithms by which to determine genetic distances among populations of the world, and thence to ascertain their biological history, or phylogeny. The HGDP fit perfectly into his overarching research program.

Pensive and downy-haired, looking very much the European aristocrat, Cavalli-Sforza is an imposing personal and intellectual figure. The force of his personality and brilliance made the HGDP newsworthy. With Walter (now *Sir* Walter) Bodmer, Cavalli-Sforza had *literally* "written the book" on human population genetics back in the 1970s. The HGDP would be the culmination of that work.

It was dubious as anthropology, however, for the program was exceedingly naïve in spite of its technological sophistication. After all, the English are not genetically distinct from the French or Irish—they have absorbed elements of the gene pools of both, in spite of their social differences. It is therefore unclear what the meaning of a phylogenetic tree of the three populations' relationships—as if they were *Drosophila melanogaster, persimilis,* and *subobscura*—would be. It is rather like asking whether archaeology is more similar to art history or to paleontology; well, it's similar to both, and the answer you get (and you can *always* get an answer from a computer) will depend on the breadth of the data you analyze and the type of analysis you make.

Cavalli-Sforza had shown in the 1960s that if you analyze the "major races," you find that EurAsians are juxtaposed against Africans in the computer-generated phylogenetic tree derived from bodily measurements; while EurAfricans were juxtaposed against Asians in the tree derived from the blood group genes. In other words, the basic division in the human species appeared to be a North-South split anatomically, but an East-West split genetically. His conclusion: the anatomy was wrong; Europeans and Africans diverge from Asians in the most fundamental division of our species. Genes don't lie.

Or do they? A different analysis of similar data by other population geneticists actually obtained the presumptively "anatomical" tree from the *genetic* data. Thus the basic structure of the tree appeared to be very sensitive to the precise computer program used to cluster the data. Perhaps, then, the conclusion was underdetermined by the data.

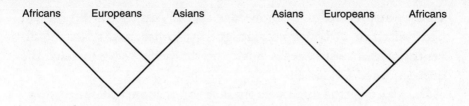

Figure 7. Are the most closely related races Europeans and Asians or Europeans and Africans? In fact, neither alternative is correct, for the African gene pool subsumes those of Asians and Europeans and therefore cannot be contrasted with them.

Even so, the model was predicated on two odd assumptions: that the human species is naturally partitioned into three equivalent units—Africans, Asians, and Europeans—and that human history can be interpreted sensibly as a series of bifurcations, as the populations come printed out of the computer. Neither assumption is valid.

Furthermore, the "mitochondrial Eve" research of the 1980s showed a pattern different from that of Cavalli-Sforza's genetic work from the 1960s. As we saw in chapter 4, African, European, and Asian are not equivalent biological categories at all, it seems, because the first subsumes the other two genetically (fig. 7).

That is a very important conclusion, because it means that asking about the relationships among the three groups is a nonsense question. As noted in chapter 4, it's like asking which two are most closely related among cows, dogs, and mammals. Since the category "mammals" includes cows and dogs, it cannot be contrasted with them. The answer you get will simply depend upon what you choose to represent your group "mammals."

Of course, those cultural categories of race are so powerful that the point has proven elusive to many geneticists. And in fact by the late 1980s, it was still eluding these geneticists as they mobilized their resources to push for the construction of a human genome diversity project (or HGDP). When the same kinds of old models (bifurcating relationships among historically constructed human groups) were used, Africans now appeared to be more different genetically from Europeans and Asians than originally thought. By 1988, Cavalli-

Sforza was prepared to acknowledge that his earlier conclusion had been mistaken. A big project designed to collect and study the diversity of the human genes might be the perfect way to solve the question once and for all.

Or, to a cynic, it might simply be a lot of money allocated to an intellectually antiquated program.

THE HUMAN GENOME *DIVERSITY* PROJECT

The proposal was published in the journal *Genome* in 1991, signed by geneticists Charles Cantor, Robert Cook-Deegan, Mary-Claire King, and Allan Wilson, in addition to Cavalli-Sforza himself. Immediately, the science media began to report on it, to help promote the project as well.

I first heard about the HGDP from Leslie Roberts, writing it up for *Science,* the most prestigious science journal in America. Roberts was interviewing people by phone for her story. In particular, she wanted my opinion on a controversy at the very top, among the organizers. Cavalli-Sforza was planning to sample the world's populations by ethnic group—in other words, to find and bleed some Hopis, some Yanomamos, some Navajos, and so on. Allan Wilson (who had supervised the mitochondrial Eve research a few years earlier) was arguing that that approach would bias the project so as to produce misleading results.

Wilson was obviously right, I explained to the journalist. As any anthropologist knows, ethnic groups are categories of human invention, not given by nature. Their boundaries are porous, their existence historically ephemeral. There are the French, but no more Franks; there are the English, but no Saxons; and Navajos, but no Anasazi. Moreover, since we have very limited access to the Franks, the Saxons, or the Anasazi (through classical physical comparisons or new methods of amplifying trace bits of DNA), we cannot really know the nature of the actual relationship of the modern group to the ancient one.

The worst mistake you can make in human biology is to confuse constructed categories with natural ones. And to load a big project with cultural categories as the overall sampling strategy would be a serious problem. First, it would make those labels appear to be genetic units; indeed, it would *make* them genetic units, which they had not been previously. Second, it would emphasize the genetic distinctions among these groups; it would force them to *be* genetically distinct by being so labeled at the outset.

What was the alternative? Wilson had suggested sampling the world by a less culturally loaded criterion: geography. Lay a grid over the world, and simply take a few people from here and a few people from there. Record the ethnic data if you like, but don't let that be the guiding variable, or else your study of human genetic diversity will be ridiculously biased toward differences between labeled categories. And we already know that those differences are quite minor. All you would do would be to magnify them artificially.

The ephemerality of those cultural labels is well attested. Wherever human populations have come into contact, they have exchanged things—technologies, ideas, DNA. Human groups—no matter where or when—trade, and where goods flow, so do genes. Those stories about the traveling salesmen are not just urban myths.

As it happened, I ended up on the cutting-room floor of that *Science* story. (I later learned that Kevin Costner had played the corpse in *The Big Chill* and had suffered the same fate, so I didn't feel so bad.) But when the story appeared, there was clearly more to it. In their zeal to drum up support for the HGDP, its advocates were making some weirdly old-fashioned assertions about human populations. For example, highlighting the "Bushmen" of southern Africa, and the Basques of the Pyrenees, Roberts wrote, "What these populations have in common is that each has been isolated and has rarely—if ever—intermixed with its neighbors. Consequently, each offers 'a window into the past,'" quoting one of the HGDP's enthusiastic population geneticists. And the reporter wasn't making it up. The original *Genomics* proposal told the scientific community:

> The populations that can tell us the most about our evolutionary past are those which have been isolated for some time, are likely to be linguistically and culturally distinct, and are often surrounded

by geographical barriers. . . . Isolated human populations contain much more informative genetic records than more recent, urban ones. Such human populations are being rapidly merged with their neighbors, however, destroying irrevocably the information needed to reconstruct our evolutionary history.

But again, as any anthropology student knows, those populations are certainly not "isolated" or pure. There are academic debates about the ethnohistories of populations for whom we have few written records, but none of them at present involves an assertion of isolation or genetic purity. And what kind of "window into the past" were they expecting to find? Did they think that the gene pool of the San "Bushmen" of the Kalahari is ancestral to the gene pool of geneticists? Just *whose* ("our") evolutionary history or evolutionary past did they expect to be illuminated by the genetics of the !Kung San or Hopi?

The fact is, human populations are—and have probably always been—porous constructs whose contemporary genetic information is very unlikely to illuminate anything but their own evolutionary history, if that. Serious scholars of human variation hadn't made assertions like that for nearly a century.

And there was more. The pitch on behalf of the diversity project asserted that time was short—because those peoples are becoming extinct. The very title of the *Science* article was "A Genetic Survey of Vanishing Peoples." In the British journal *Nature,* biologist Jared Diamond stumped for the project with the impassioned plea that "so little time remains before it will be too late."

But wait a minute. If people are "vanishing," what good does it do them to save their genes?

These scientists were trying to approach indigenous peoples whose lands had been stolen, lifeways eradicated, and people exterminated, at the hands of the very civilization the scientists represented.

And now they wanted blood.

The cultural assumptions on the part of the geneticists were stunning: for people whose lifeways and whose very existence are jeopardized by insensitive exploitation, the preservation of their *cells* bespoke a very cynical hereditarianism. The geneticists apparently expected to look people in the eye and tell them that their DNA was more valuable than their customs, their land, their traditions, and their lives. Imagine the scientists' surprise when those people failed to share their priorities!

And why is the fluidity of human populations ("merg[ing] with their neighbors") such a bad thing? Did they think it has only just begun? In fact, many of the "vanishing" populations were only vanishing in a peculiarly narrow, genetical sense. Many were actually bouncing back demographically and culturally, through a revival of traditions, immigration, and the general flux of population histories. Their genetics were changing, but why would geneticists demand them to be frozen in time?

Worse still, when they were approached by geneticists interested in studying the gene pool of African Americans and Latinos, the leaders of the HGDP demurred. Those were admixed populations and thus not interesting to them. The goal of the project was apparently to reconstruct an imaginary human gene pool of the year 1500, not to study the real human gene pool of the year 2000.

Objections to the project were slow to be voiced because the geneticists had succeeded in building up a favorable publicity machine in the science media. Not only was this a big anthropology project conceived and designed without input from anthropologists, but neither was there input from, or contact with, the people they hoped to study. It thus appeared to be a classically colonialist undertaking. Geneticists regarded the indigenous peoples as objects of analysis; the scientists were talking *about* them, but not *to* them. From a standpoint of unequal power relations, they would coax them to provide some of their blood (or perhaps get someone else to do the dirty work), and then find out who they were closely related to, back home comfortably in Palo Alto.

It was a throwback to the anthropology of a much earlier generation.

With financial assistance from major scientific organizations, the geneticists began to hold formal meetings to discuss the scientific issues. In July 1992, they convened at Stanford to debate the issues of theoretical population genetics that would be central to the project's aims. Later that year, they convened again at Penn State to decide which populations should be targeted, and came up with a list of about 500, which ultimately became an official list of 722 populations. A third news feature in *Science* again reported the project's two articulated goals: "how populations are related" and "to probe human evolutionary history." Anthropologists were now being invited to help, but it was clear that their participation was not for the project's design, but simply to help operationalize it.

OBJECTIONS TO THE PROJECT

Times, however, had changed. As anthropology had matured as a discipline in the 1950s and 1960s, the scholars who were dependent upon indigenous peoples for their academic livelihoods saw themselves increasingly as responsible *to* those people, acting as voices for people who had little or no voice. Anthropology worked at getting their stories told and helping develop their voices. Perhaps it was a bit paternalistic, but that was better than colonialistic, and it was certainly motivated by more noble goals. And now many of those people had voices, and had some measure of political clout. It had even begun to backfire, as Native Americans became empowered by the Native American Graves Protection and Repatriation Act (NAGPRA) of 1990, reclaiming skeletal materials and artifacts that had been mainstays of anthropological studies for decades.

And these same "isolated" and "vanishing" peoples now began to speak out about the Human Genome Diversity Project.

In early 1993, a watchdog organization concerned with the rights of indigenous peoples, the Rural Advancement Foundation International (RAFI), alerted representatives of several populations about the well-hyped project. They saw things a bit differently than the scientists did. They recalled the free use of local knowledge of plants by major First World corporations, which then developed hybrid strains, patented them, and sold them back to the people who made the profits possible in the first place without sharing in them. They recalled a long history of relations with powerful Euro-American institutions, ranging from being exploited to being exterminated, but usually ending in their being impoverished and socially marginalized.

By late 1993, e-mail had linked up the "isolated" populations and their activist spokespeople into a strident anti-HGDP network. Open letters to the HGDP bioethicist and lawyer Henry Greely from the South-and-Meso American Indian Information Center (SAIIC), and from the Onandaga Council of Chiefs to Jonathan Friedlaender of the National Science Foundation, a supporter of the HGDP, were widely circulated in the academic community. In December, a World Council of Indigenous Peoples in Guatemala repudiated the HGDP.

As a reaction to the first wave of anthropological criticisms, the HGDP attempted to shift the major popularized goals of the project away from human microphylogeny and toward biomedical benefits that might accrue from a study of genetic diversity, such as isolating and thereby treating genetic dispositions to hypertension and diabetes. This was difficult to sustain, however, since the project had only planned to collect blood, not biomedical histories, or even general phenotypes. Without a knowledge of *who actually had diabetes*, it would be virtually impossible to collect any useful medical genetic information on it. For a genetic study of height, you have to know how tall the subjects who donated the DNA being studied actually were.

The project's goal was to record the ethnic group of each DNA sample, since that was what its organizers needed to know, given their research interests. Thus any researcher who wanted such a sample would receive a "Hopi, female" or "Efe pygmy, male." Unless diabetes (i.e., medical history) or height were serendipitously recorded, that information would simply not be available for a researcher interested in such matters. There was no plan for the systematic recording of medical history or outward characteristics (phenotypes); the HGDP was being conceived simply for molecular population genetics, for researchers with a biohistorical interest, like Cavalli-Sforza's, nothing more.

It was already known that the biohistory of human populations is difficult to extract from genetic differences. Different populations might be similar to one another because they diverged from one another recently, or because they came into genetic contact recently. One could certainly generate a bifurcating tree from genetic data, but it was difficult to know exactly what the tree might mean. That tree, however, was the HGDP's raison d'être, even though representatives of indigenous peoples themselves had little interest in it.

The scientists were still meeting among themselves to try and spin-doctor the situation. They began to appreciate that the concept of "informed consent," vital to modern bioethical protocols, might be difficult to obtain when seeking to extract genetic materials from people who did not share their conceptions of the body, the blood, and hereditary continuity. As early as 1922, the anthropologist Carleton Coon (see chapter 4) had collected blood samples from the Rif in Morocco for study in the United States. Coon is not remembered as being among the most "politically correct" anthropologists of his generation, but he knew it was no small matter to coax people to surrender their bodily fluids. "Blood-letting for blood-group analysis falls into the class of blood-letting in general, and evokes the whole ideology of blood-brotherhood, the fear of injury by contagious magic, and that of ritual condemnation based on the analogy of menstruation," he observed in his popular book *The Story of Man* (1954).

And that was when the distinction between consent and coercion was blurry and it didn't really matter whether the participants in a scientific project understood what you were doing to them. It did now. The very first article of the Nuremberg Code, the foundation of modern (i.e., post-Nazi) bioethics, reads:

> The voluntary consent of the human subject is absolutely essential. This means that the person involved should have legal capacity to give consent; should be so situated as to be able to exercise free power of choice, without the intervention of any element of force, fraud, deceit, duress, over-reaching, or other ulterior form of constraint or coercion; and should have sufficient knowledge and comprehension of the subject matter as to enable him to make an understanding and enlightened decision. The latter element requires that before the acceptance of an affirmative decision by the experimental subject there should be made known to him nature, duration, and purpose of the experiment; the method and means

by which it is to be conducted. . . . The duty and responsibility for ascertaining the quality of the consent rests with each individual who initiates, directs, or engages in the experiment.

That was, of course, intended to cover medical experimentation on citizens of the First World. Drawing blood for genetic analysis—although medically rather benign—would certainly fall under it; and it should be applicable to the rest of the world's inhabitants as well. In practice, however, it would obviously be very difficult to secure such consent. To do so, someone would have to discuss cell biology and molecular genetics with people to whom the scientific mind-set involved was alien. And it's hard enough to explain those concepts to elite First World college students!

The population geneticists had gone to bed with a nice idea, the amassing of a large repository of data on the populations of the world, and had awakened on the cutting edge of contemporary anthropology and bioethics. And they were entirely unprepared for it.

In terms that recalled Carleton Coon's almost pathetic attempts to deny responsibility for the social and political implications of his racist anthropological publications in the 1960s, the HGDP indignantly castigated their opposition for politicizing their science. An exasperated HGDP geneticist complained to *Discover* magazine in November 1994, "We're scientists, not politicians."

As if opening the veins of the indigenous peoples of the world might not constitute a significantly political act!

And this was over thirty years after Coon's *The Origin of Races*. Either these scientists were just too dumb to be entrusted with such a sensitive project, or they were being disingenuous when they blamed "the other side" for politicizing it. By now anthropologists were becoming vocal critics of the project as well.

RACE AND THE PROJECT

The year 1995 began auspiciously for the HGDP. Cavalli-Sforza had just published a magnum opus called *The History and Geography of*

Human Genes, and *Time* ran a feature on him, on the book, and, of course, on the HGDP. Once again, though, things began to backfire on them. Taking another tack—that the HGDP would combat racism (and who could object to that?)—*Time* represented the HGDP as a counterblast to *The Bell Curve.*

But that was the equivalent of calling *Gone With the Wind* a refutation of *Moby-Dick.* The HGDP simply had nothing to do with the thesis of *The Bell Curve.*

However, the leading proponents of the HGDP all had impeccably liberal credentials dating from the 1970s, when the psychologist Arthur Jensen and the physicist William Shockley were reinventing the inherent inferiority of the American lower classes. Luca Cavalli-Sforza and his colleague Marcus Feldman had both been strident opponents of the folk heredity of that era.

So this became a new selling-point for the HGDP. It was now going to make the world a better place, because it was going to show that there are no discrete races and thereby undermine racism. But, of course, we already knew that there are no discrete races (Ashley Montagu had said that in the *Journal of Heredity* as early as 1941), and there doesn't seem to be much sense in undertaking a big scientific project to find out something we already know.

Moreover, that goal assumes that the problem of racism lies in animosities between distinct natural groups, so that denying the existence of such natural groups would implicitly deny racism a basis. But, again, we already know that races are cultural constructions and that racism exists independently of the naturalness of races. The people who hate one another the most are generally very similar biologically: think of the Irish and English, Hutu and Tutsi, Huron and Iroquois.

Racism is thus independent of the existence of large natural groups of people, or races. What could the Human Genome Diversity Project tell Bosnian Serbs and Muslims about each other that they don't already know, or that they would care about?

Once again, the HGDP was talking through its hat. It had nothing to contribute to contemporary discourse on racism. And nothing revolutionary to tell us about the major features of human variation.

But things quickly worsened, for *Time* ran a color figure from *The History and Geography of Human Genes,* in which each of the non-existent human races actually came color-coded: Africans yellow, Mongoloids blue, Caucasoids green, and Australians red.

Quite literally!

It was the old essentialist fallacy of Linnaeus, except now with different colors and computerized.

And just for good measure, *Time* told its readers, "All Europeans are thought to be a hybrid population, with 65% Asian and 35% African genes."

To say that statement was "wrong" would imply that we really know the proportion to be, say, 80:20, rather than 65:35. To call it wrong would be a massive understatement. It is *less* than wrong; it is not *even* wrong.

There is no reason at all to think that there was ever a time when there were only Africans and Asians who interbred to create Europeans. There is no reason to think that Africans and Asians are purer or more homogeneous than Europeans, or even that all three of these geographical categories are equivalent biologically (in fact, as we have noted, Asians and Europeans appear to constitute genetic subsets of Africans). So that statement reflects a great deal of misconception about human biology, diversity, and history. Was it a misquotation? Did *Time* just make it up? No—the HGDP's advocates had published a scientific paper in 1991 purporting to show it. "[A]ncestral Europeans are estimated to be an admixture of 65% ancestral Chinese and 35% ancestral Africans," they wrote in the *Proceedings of the National Academy of Sciences.*

That kind of thinking has a venerable history in Western thought. To an earlier generation, dividing humans into three continental types harmonized well with a mythic history that saw humans as descended from Noah's three sons. Although the far reaches of the continents were unknown to them, by fanciful extrapolation from the Book of Genesis it was argued that the ancient Hebrews had ascribed North Africans to the lineage of Ham, central and southern Europeans to the lineage of Japheth, and West Asians (including themselves) to the lineage of Shem, "after their families, after their tongues, in their lands, in their nations" (Gen. 10:20). And this origin myth ultimately became the framework used by early nineteenth-century anthropologists to understand racial variation: the most divergent peoples became the purest, most primordial peoples, and they could be found in West Africa, East Asia, and northern Europe. But serious scientists had abandoned that kind of thinking decades earlier, until the new genetic work reimposed it on their data.

In fact, the geneticists' "Asians" were just a small sample of people from China, now living in the San Francisco Bay Area, and their "Africans" were two populations of Pygmies. This was some of the most crude, culturally loaded, pseudoscientific prattle imaginable in the contemporary genetics literature, and it was now exposed at the very intellectual core of the HGDP.

Nevertheless, the HGDP had managed to get a seed grant of $500,000 from the National Science Foundation, as well as seven-figure grants from other private foundations. However bizarre the research seemed to be, it was making an impression.

By year's end, however, there was yet another scandal brewing.

THE CUTTING EDGE OF BIOETHICS

The Hagahai are a group of highland New Guineans who live, as most human groups do, under constant exposure to certain pathogens. And they have, as most human groups also have, their own suite of minor genetic distinctions. In this case, some of the Hagahai villagers posses a unique form of a human leukemia virus, which is benign, unlike its closely related viral sequences in the rest of the world.

Carol Jenkins, an anthropologist working with the Hagahai, had retrieved their blood, had tried to explain cells, had tried to explain the Western concept of patenting, and (trying to act in their interests) had brokered a deal whereby the Hagahai would share in the profits, if any, that a patent on the cell line containing the benign leukemia virus might bring.

Jenkins had nothing to do with the HGDP, except that she was doing on a small scale, as many scientists have been, what the HGDP proposed to do on a centralized, grand scale. And in the spotlight that the HGDP brought to this field, Jenkins happened to be in the proverbially wrong place at the wrong time.

On March 14, 1995, a patent was issued to the National Institutes of Health (NIH) for the Hagahai cell line. This, of course, was nearly precisely the scenario RAFI and other indigenous-rights groups had predicted. And where the HGDP had maintained all along that there were no financial interests at stake in the collection of blood from diverse peoples, clearly there were indeed.

Moreover, all the purple prose about ultimate significance of DNA in understanding human life, formerly used to promote the Human Genome Project in the late 1980s, was now thrown back in the faces of the Diversity Project. RAFI posted the story on the Internet. "The United States Government has issued itself a patent on a foreign citizen," it proclaimed. "[The Hagahai] now find their genetic material—the very core of their physical identity—the property of the United States Government."

Where formerly geneticists had gushed that the genetic information in one's cells contained "The Book of Life" or "total knowledge of the human organism," they now scrambled to reassure the public that they were not patenting a person, or trying to gain ownership of a human essence, but "just" a cell line.

As it happened, patents had also been filed for cell lines derived from a Guaymi woman from Panama and a Solomon Islander. So this was not an isolated incident, but a serious issue, indeed, the leading edge of bioethics. Do people have a financial interest in the profits derived from their own bodies? Does the act of donating a blood sample constitute a blank check for all future biotechnological research, however lucrative?

Once again, unfortunately, the Human Genome Diversity Project was unprepared to discuss these issues. Its leaders indignantly denied that they had anything to do with the patents in question, which was true, but irrelevant—for the patenting issue certainly had major implications for the project. And they somewhat more sheepishly denied that cells cultured in vitro were the equivalent of a person's life essence, which again was true, but difficult to reconcile with the popular promotional spins the molecular biologists had been disseminating on behalf of their craft for the past decade.

Indigenous peoples and their advocates had good reason to be worried. On July 9, 1990, the California Supreme Court had set a landmark precedent in the area of biotechnology when it ruled that the patent on a cell line known as "Mo-Cell," derived from the spleen of a cancer patient named John Moore, belonged to the doctors who had "created" it. Moore, who had refused to sign a consent form, nevertheless still did not own any interest in the products derived from his own body. And he was a middle-class white guy from Seattle.

Imagine how vulnerable the rest of the world was to rapacious biomedical capitalism!

I heard a talk once by an expert in patenting, who explained to a genetics conference in Milwaukee that American law favors the scientist patenting the cells, not the source of the cells, because the scientist has put the labor into the cell line, isolating and characterizing the DNA. This being the case, and knowing of the HGDP's interest specifically in indigenous people and their presumptively exotic gene pools, I asked a question from the audience.

I was sitting at a table with some Native American friends interested in the HGDP, as after all the HGDP was interested in them (or at least, in their blood). So I asked the speaker, "Given what you've told us about patent law, I understand why a geneticist would want to collect blood from Indians—there is a chance, however small, that they could make some big money. My question is, Why would an Indian wish to give blood to a geneticist?"

The speaker thought for a second, and gave an answer that inadvertently sent our table into paroxysms of laughter.

She said, "Umm . . . , altruism?"

THE HGDP'S DEATH THROES

At a symposium on the HGDP at the International Congress of Bioethics in San Francisco in 1996, geneticist and bioethicist Benno Muller-Hill addressed a question to the HGDP's leaders. Given that we are going to find some genes someday that affect some aspects of intelligence, and that we are going to find them to be unevenly distributed across populations, Muller-Hill asked, as custodians of the human gene pool, how will the HGDP deal with that information?

They had no idea; apparently, it hadn't occurred to them. The answer they gave was simply to deny that there might be genes for intelligence. The bioethics audience was, in Dorothy Parker's famous coinage, underwhelmed.

That was the beginning of the end. And the end came swiftly.

The final blow came less than a year later, in October 1997, when a blue-ribbon panel convened by the National Academy of Sciences and the National Research Council released its findings about the Human Genome Diversity Project.

Of course, it said, the study of human genetic diversity is an important issue. But the HGDP wasn't going about it properly. In the first place, observed the panel's report, "the precise nature of the proposed HGDP was elusive; different participants . . . had quite different perceptions of the intent of the project."

In other words, seven years and several million dollars into the project, its spokespeople couldn't even tell the panel what it was for!

Their major concern was that HGDP geneticists had not paid sufficient concern to the bioethical issues they were raising. Probably they were influenced by the chillingly naïve testimony of a geneticist who said on the record that he saw no ethical problems in collecting blood. Just go to an Indian reservation with some $5 bills, he proposed; there are so many desperate people there who will sell you anything for five bucks, you'll have no problem getting samples.

To him, it was a simple market economy. You go up to some needy people and offer them a mess of pottage for the rights to a commodity you want and they have, and that you will be able to advance your career from, and possibly even become rich from.

In spite of the biblical precedent, there is a major bioethical problem: The participant doesn't really have a choice. This is not a case of voluntary informed consent; it's a case of taking crass advantage of a tragic situation.

Hardly the kind of thing that casts science in a favorable light.

Faced with a rebuff from the National Research Council's panel, the HGDP's supporters tried to spin-doctor the report as best they could. After four gung-ho write-ups in *Science* in the 1990s, its news department came up with a story titled "NRC OKs Long-Delayed Survey of Human Genome Diversity," somehow managing to find the opposite message in the report. *Science's* most bizarre article a year later actually attempted to credit the HGDP geneticists with discovering that "genetic diversity appears to be a continuum, with no clear breaks delineating racial groups"—as if anthropologists hadn't been saying that for decades, and ignoring the color-coded racial maps the HGDP geneticists had actually been publishing!

REQUIESCAT IN PACE

The Human Genome Diversity Project raised many important issues that had been previously hidden from view; things that geneticists had been getting away with, for better or worse, on a small scale. The public campaign for big money brought with it the glare of a far more intense scrutiny than the geneticists had been accustomed to receiving. It suddenly brought out questions that had not been asked, questions about science and people.

Suppose, for example, you are interested in the possibility of genetic causes of alcoholism. You know that rats can be bred with genes that compel them to prefer to drink greater and greater quantities of alcohol. Maybe there are humans like that, too. You need a human population in which there is rampant alcoholism, and a population that permits you to take their blood for study.

You go to the Navajos, and explain how if your studies work out, you might be able to diagnose people at risk for alcoholism and maybe cure it someday with gene therapy.

But you also know that only a tiny component of alcoholism on the Navajo reservation—if that!—is due to genetics. Navajo alcoholism is a terrible tragedy and an important social problem, but it is due in the main to nongenetic causes.

So why is your work getting funded? Why isn't that money being spent on the overwhelming side of the problem that is *not* genetic?

If the Navajos themselves had any input in determining the priorities for research, or access to the resources, they might well choose to develop programs that would approach the problem differently and get at its nongenetic root causes. In fact, if you bothered to explain to them how small a component of alcoholism genes really account for, it's conceivable that the Navajos might not consider genetic research into the subject important enough for them to participate in.

Rather, they might insist that you take your syringes instead to Boston on St. Patrick's Day.

But that is the crux of the issue for geneticists at the millennium: Who decides what genetic problems are important? Traditionally, it has been the scientists defining the research questions, with indigenous peoples as passive pincushions. With the empowerment of indigenous peoples over the past few decades, that is no longer tenable.

Indigenous peoples are increasingly weighing the merits of scientific questions themselves and making scientists justify themselves to a much greater extent—and the less scientists acknowledge that, the less science they will be permitted to do. This is a global political reality in the twenty-first century, and the fact that the HGDP's leaders did not recognize it is another testimony to their naïveté or insensitivity.

The issue is not the study of human genetic diversity per se but the ancillary questions of bioethics, finance, discrimination, confidentiality, essentialism, and choice. The ones the HGDP refused to address.

Decentralized, unregulated collection of genetic materials from the peoples of the world is still going on. There ought to be some kind of centralized human population genetics project, a fusion of anthropological problems and genetic technologies. But it will have to use the conceptual apparatus of modern anthropology. It will have to deal with the fact that the scientific study of people is invariably political; it's just not like the scientific study of clams. It will have to address a broader suite of questions than just microphylogenetic inferences from trees based on genetic distance. And it will have to address bioethical issues proactively and not merely deal with them as they happen to come up—by which time it will be too late.

I started out skeptical of the HGDP and evolved into an enemy. The scientists in charge had no interest in acknowledging errors, in revising their public positions, or in even learning about the relevant cognate material. For them, it was a simple matter of getting lots of money, getting blood, and finding out who was related to whom. It was never clear they could do the third; they belatedly and grudgingly acknowledged the limitations of the second; and, as a consequence, they are being denied the first.

If there is a special section of hell where scientists spend eternity after ruining a good idea and leaving others to mop up the mess they've made, that's where you'll find the stalwarts of the Human Genome Diversity Project. Perhaps they'll be plotting to study the genetic diversity of the Damned.

Now that might be interesting.

Ten

IDENTITY AND DESCENT

THERE IS A QUESTION THAT I am frequently asked, usually by wide-eyed students who have already taken their science requirements, often in preparation for medical school, and who are now settling in for a semester of biological anthropology, an elective that looks at least relevant to their scientific interests. It comes up when I arrive at the topic of molecular anthropology, and of the genetic similarities between apes and humans.

"Given the genetic similarities between chimp and human," begins the student, "could they interbreed? Could we make a human-chimp hybrid?"

Now, there are a lot of self-defeating ways to deal with that question. One of them is to take the attitude that there are things we are not "meant" to know. But such arguments will carry little weight, since the source of their moral authority is not clear. Who says we aren't meant to know? Who is the custodian of what we are and aren't meant to know? Weren't we meant to know about the immune system, so that we could be spared smallpox and polio? Weren't we meant to know about the internal combustion engine, laser beams, and radio waves?

How, then, can we say that we *aren't* meant to know about chimps and humans?

A second poor way to deal with that question is to look heavenward and intone that we shouldn't "play God" by "violating the natural order." But haven't we been playing God and violating the natural order since we began growing our own food and domesticating animals? Those acts took plants and animals that had formerly been wholly in the domain of nature and placed them under human control and created genetic entities that had never existed before. God's domain has never been particularly well demarcated in relation to science, and it is certainly no great homage to the Deity to define it as simply what we cannot do or conceptualize as scientists at any particular point in time.

A third, and rather more popular, fruitless way to deal with the question of an ape-human hybrid is to wag one's finger and proclaim that it would be immoral and unethical. The problem with this answer is that there is a follow-up question that goes, "What would be immoral and unethical about it?" At this point, the average genetically trained scientist sighs reflectively and explains patiently to the inquisitive student that they simply couldn't get funding for such a research project.

Hopefully true, but decidedly a non sequitur.

The "chimp-human hybrid question" is indeed a thorny one, but it brings us back to the core of molecular anthropology, the fusion of humanistic values and scientific technologies.

Could we make a human-chimp hybrid? I used to answer that question facetiously: "It's already been done; how else can you account for Pat Buchanan?" Har-de-har-har.

But I take the question a lot more seriously now, because I think it overlays a lot of important things that we take for granted in science education these days, and which need to reconsidered.

The answer I give nowadays is this: Your guess is as good as mine, probably better. But let's say you do the experiment. What are you going to do with the baby? How are you going to raise him/her/it? As an ape? As a human? You are talking about producing an organism outside of any natural society. *What do you think your responsibilities are to that baby?*

Are you going to raise him/her/it in isolation from the rest of the world, which would be psychologically destructive to an ape *or* a human; or announce it to the world, and allow *People Magazine* in to film the offspring's life? What kind of an identity will you help him/her/it achieve? Bi-*racial* children have a hard enough time.

Assuming you have produced a sentient articulate being, how will you answer the questions: What am I? Why don't I fit in? Why is no one else like me?

Budding geneticists learn to think technologically. The current generation are also being forced to think about their responsibilities and obligations to their subjects and about accountability. That is the heart of molecular anthropology.

"Could it be done" is not simply a biochemical question, it is a social and ethical question. And let me make it clear that by "ethical," I don't mean to suggest that there are some things we shouldn't know, because I think we should strive to know everything. By "ethical" I mean it's a question that involves acknowledging responsibilities on the part of the researcher to the objects of analysis. Doing research on humans may be technological, but doing it *right* is a social and ethical matter.

In other words, the technical aspects of that experiment are the easiest. What to do with the child once you've created it is a much harder question.

That hybrid infant would be in a perpetual state of intermediacy, only half-person.

What would it be? Its human part would certainly want to know, because one of the most quintessentially human attributes is the desire to situate ourselves in the social universe. We establish our identity through systems of kinship; we decide *who* we are by learning *what* we are.

WHO AND WHAT

Kinship is familiarity; kinship is orientation. That is why we build so extensively on the metaphors of kinship—"we are family" as a baseball team's slogan; "fraternal" lodges; "Mother" Teresa; "Uncle" Sam; "sorority" houses.

A classic ethnocentrism is to suppose that our culture's kin terms are rooted in nature or genetic relations, and that other cultures' kinship systems are not. But the diverse relations subsumed by the term "aunt" (see chapter 2) easily give the lie to that. Moreover, our kinship system does not represent genetic relationships among lineal kin straightforwardly: I am a mitochondrial clone of my mother's mother, but she bears the same relation and kin term as my father's mother, whose mtDNA is not closely related to mine. Likewise, my Y chromosome is my father's father's; and my mother's father did not give me a Y chromosome, although both are my grandfathers.

In other words, kinship is the imposition of order upon a complex matrix of social relationships. There are many kinds of possible order, but no kinship system actually represents genetic relationships accurately. In our culture, however, we look to science for authority. It's reassuring to know that our diets are scientifically approved, that our medicines are scientifically chosen. It's reassuring to know our social prejudices are scientifically validated (and not based on the whims of selfish, greedy, evil ancestors—perish the thought!). We look to science for answers the way earlier generations looked to their magi (which, of course, is translated as either "wise men" or "magicians").

But what about when our ideas about our place in the natural order, who and what we are, are impinged upon by science?

Science gives us authoritative new ideas about kinship, which force us to reconceptualize our place in the order of things, which is by that very fact disorienting. But it doesn't stick around to explain it to us, to reintegrate us, to give new meaning to our existence. That's the problem with Darwinian theory, of course. It tells us our ancestors were kin to apes, the products of eons of ordinary biological processes of survival and reproduction, and not merely zapped into existence in the Garden of Eden, *but it doesn't tell us what that means or what to do about it.*

It just walks away from the wreckage.

And the question of who and what you are is not trivial. At a symposium in 1996 on the Human Genome Diversity Project, I found myself sitting next to Debra Harry, a Northern Paiute activist against "bio-colonialism" who took a postgraduate fellowship to learn genetics and quickly became a major thorn in the side of the genetic bio-pirates represented by the HGDP. A spokesman for the project came to the podium and explained patiently its rationale—"We're going to tell these people who they really are." What presumption! As if they needed science to tell them! Debra took the floor next and responded, "I know who I really am—shall I tell you who you really are?"

CLONING

Perhaps the best contemporary example of science impinging on kinship is in the cloning controversy. It surfaced briefly in 1978, when a writer named David Rorvick published a book entitled *In His Image: The Cloning of a Man,* about a reclusive millionaire who buys himself some geneticists and reproductive biologists and contrives to have himself cloned. It was a novel, but Rorvick misrepresented the story as nonfiction, setting off a ferocious controversy about whether or not it could be true.

Much of contemporary thought about cloning was shaped by popular fiction, notably Woody Allen's movie *Sleeper* and Ira Levin's novel *The Boys from Brazil* (and the film based on it), both of which date from the mid-1970s. Each centers on a plot to clone an evil dictator, linking cloning technology to its application.

But the debates over cloning died down with the discovery that Rorvick's story was fiction, as were all the other cloning scenarios. Scientists vowed it couldn't happen, and so there was no need to worry about it.

Then came Dolly, the cloned ewe, in 1998. Dolly suddenly ushered in a new wave of millennial technophobia, calling forth scenarios in which father and grandfather could be the same person, in which fatherhood per se might not even exist, in which generations would be blurred, in which the wealthy would clone themselves to perpetuate the exploitation of the poor, and tycoons would keep twins or triplets of themselves on ice to use for spare parts.

Human cloning was immediately condemned by people who make a living condemning things, but I confess I never saw much horror in it. Hitler was a product of his age—abstracted from his time and place, he would be a nobody. A hundred clones of him would be a hundred nobodies. The fear of cloning Hitler is very much an artifact of our cultural belief in the innateness of prominence. Does anyone really think that a Hitler clone at Woodstock would have formed an army of hippies to take over the U.S. government?

Nor am I trembling at the prospect of cloning as a tool of the wealthy. It seems to me that a Brad Pitt *raised* by Jennifer Aniston would have to come out somewhat differently than a Brad Pitt *married* to Jennifer Aniston.

And I see no reason to think that a world with six Bill Gateses, five Ted Turners, four David Geffens, and three Donald Trumps would be manifestly inferior to a world with only one of each.

The problem with cloning is not the threat it might pose to the world, but to the rights of the clone. The clone would be nothing more than an identical twin, but it would be a different age from its twin. That means it would have an idea of its genetic liabilities and assets in advance of their development, which would inevitably place constraints on its right to an open future, as some bioethicists have noted. Imagine the burden of being a genetic replica of Michael Jordan, or Itzhak Perlman, or Neil Simon. What choices would you really have about your own life trajectory? As a sentient and functional member of our species, you would be entitled to such choices, but as a clone of a prominent preexisting member, your choice would be curtailed—a clone of Neil Simon would inevitably be pushed toward the theater, regardless of how he felt about it.

In other words, the problem of cloning is that it would create a situation in which a person had an already partly formed identity, a situation that of course never arises with the "natural" analog of clones, identical twins. They explore and create their identities together. The most famous twins of all, Chang and Eng—the eponymous Siamese Twins—developed separate identities and personalities while sharing all their DNA and many body parts, notes biologist Paul Ehrlich.

Cloning thus technologically impinges on a crucial human right, that of self-discovery. The right to an ancestry, to a lineage, and to an independent identity in relation to your ancestors and relatives. That can't be taken for granted.

Furthermore, this is a path that shouldn't be trodden. When science begins to impinge on people's ideas about *who* they are and *what* they are, it encroaches on humanistic concerns and issues. The issue of cloning has little to do with spare body parts in the closet, or an army of Jeffrey Dahmers, and everything to do with having a feeling of confidence in where you came from, how you fit in, and what you can strive for in your life. Science can participate constructively, but scientists have to realize that the hard part isn't the technical part; it's the social and cultural part.

KENNEWICK MAN: THE FRENCH AND INDIAN WAR

Frederika Kaestle now teaches at Indiana University, having been an outstanding undergraduate student of mine at Yale in the 1980s. Her undergraduate thesis was on the genetics of the apes, and she went on to do a doctoral thesis at the University of California–Davis, on the biohistorical relationships of western Indians. Kaestle was in on the ground floor of the development of a new series of techniques for studying those relationships—ancient DNA. Since bones contain cells, and cells don't vaporize immediately upon the death of the organism containing them, sometimes bits of DNA can be extracted from the bones of long-dead animals. The most extreme claims (DNA hundreds of millions of years old) appear now to have been as fictitious as *Jurassic Park*, but it is not terribly uncommon for researchers now to be able to extract a little genetic material reliably from bones or carcasses thousands of years old. They've even gotten a tiny bit of DNA from Neandertals and have learned that they were very similar to, but a bit different from, modern people (as if we didn't already know that).

But Rika Kaestle wasn't working on Neandertals. She was working on Indians. As a tiny part of her thesis, Kaestle was asked to analyze the DNA from a new specimen that had come to be known as "Kennewick Man." Unfortunately, her study was interrupted, because the specimen was confiscated by the federal government. It wasn't as if she was trying to clone Hitler, as in *The Boys from Brazil*, but somehow her work had quickly become as powerful and dangerous as any genetic study could be, and it was shut down.

Why? How could the study of ancient bones be so threatening? To get at the answer, which is at root a story of the abuse of scientific authority, we must reconsider the history of anthropology. It is now popularly perceived, albeit somewhat vacuously, as a "politically correct" field, teaching cultural relativism, the respect for exotic and diverse lifeways. And that is undeniably a part of it. But what is entailed by such respect? Standards of respect themselves evolve, and even the most open-minded of anthropologists a hundred years ago would have difficulty with contemporary standards.

How does one go about studying the history of peoples, and of their civilizations and cultures? The obvious answer is, you dig it up— that is the heart of archaeology, itself a part of anthropology. And what you dig up is often a cultural artifact, such as a scrap of pottery, or a tool or weapon, or, commonly, a bone, perhaps of an animal eaten, perhaps the remains of one of the people you're studying. Macabre business, to be sure. But the study of human bones can in some contexts be very rewarding—it can assist police work by diagnosing the age and sex of a long-dead victim, for example. It can tell you about the traumas that befell particular people, or even fossil people, like Neandertals. It can tell you about their health, their average lifespan, and even, perhaps, something about who they were and where they came from.

Now, where did all this detailed knowledge come from? The simple answer: from studying skeletons—how and under what circumstances they come to differ from one another.

But whose skeletons? Certainly you're not going to dig up your own grandmother to study her; that would be disrespectful and sacrilegious, in addition to being grisly. If only there were other bones around to study, bones that weren't yours, your family's, or those of anyone like you, or that anyone cared about . . .

And therein lies the paradoxical beginning of modern physical anthropology in the nineteenth century, with the collection and study of the bones of dead Indians. Thousands of bones of Indians from all ages, the results of controlled archaeological excavations and of simple grave-robbing, have formed the basis of much of our knowledge of human skeletal variation, and they sit now in the dank recesses of museums.

One of the major roles of "cultural relativism" is as a conscience of the industrial world. Not merely pointing out the diversity of behavioral traditions for the sake of the titillation of exoticism, but compelling students, or just thoughtful citizens, to appreciate the precariousness of history in light of the privileges of power. You might pause to think about how the course of your life would run if indigenous Americans had conquered and colonized Europe, rather than vice versa, and what you might then hope for in the way of toleration or respect for your beliefs and behavioral traditions. ("It could never have happened that way" is the classic response, based on the insecurities evoked by the thought of being unprivileged oneself, but we know of no innate flaw in the colonized peoples that would make this scenario impossible.)

A product of the consciousnesses raised by a century of anthropology in America is the legislation known as NAGPRA, the Native American Graves Protection and Repatriation Act (25 U.S.C. 3001–3013), introduced by Congressman Mo Udall (D-Ariz.) and ratified in 1990. NAGPRA has two major injunctions: first, Native American cultural or biological materials stored in museums must be catalogued, appropriate tribal representatives must be contacted, and the materials must be surrendered if the tribes want them; and second, any artifacts or remains found on federal lands must now be treated as if owned by the Indians, and you have to get their permission to study them.

Its critics called NAGPRA "anti-science," but it isn't science legislation, it's human rights legislation. It's about setting limits on scientists, who can no longer do just anything they damn please. NAGPRA is as anti-science as the Nuremberg Code. Sure, it might be interesting to know just what the tolerance of pain of the human body is, or the precise amount of X-rays it takes to sterilize or kill a person, or what syphilis would ultimately do to someone who was left untreated, but those are scientific questions that cannot be directly studied, by virtue of the excesses of previous generations of scientists. Likewise, NAGPRA is an acknowledgment that those old bones are somebody's; they are meaningful to people in nonscientific ways, and those ways are important to acknowledge, and must be weighed against the science.

NAGPRA takes science at its word: that it is for the enrichment of people's lives. In any case, it's not to make them miserable and insecure.

KENNE-HYPE

The *Tri-City Herald* in Washington State ran a small story in its edition of July 29, 1996, to the effect that a couple of students watching a boat race from the shores of the Columbia River had found a skull and a few other bone fragments. They turned the remains over to local police in Kennewick, Washington. Nothing was known of the age or origin of the skull, and it was being studied by a local crime lab.

Late that afternoon, the county coroner called James Chatters, who runs a local forensic anthropology business. Chatters recognized that the skull wasn't from the victim of a recent crime; it was old. It was possibly the skull of an early settler, although the back teeth were exceedingly worn down, like those of Indians. Chatters applied for a federal permit to look for more bones, and received it on July 31, retroactive to July 28, the day the skull was first discovered. They found most of the skeleton, in bits and pieces ultimately numbering about 350.

The skeleton seemed to be from a large adult man, a bit less than six feet tall, with a constellation of facial characteristics not typically Indian. On the other hand, he had a very archaic stone point, two inches long, embedded in his hip. Chatters had a CAT-scan performed on the pelvis, to see the precise features of the spear tip, and it was clearly a kind only in use thousands of years before any European (Lewis and Clark were the first) had ever set foot in the region.

According to Chatters, "We either had an ancient individual with physical characteristics unlike later native peoples, or a trapper/explorer who'd had difficulties with 'stone age' peoples during his travels." To resolve this ambiguity, therefore, he sent part of the bone just below the left pinkie to a radiocarbon-dating laboratory on August 5.

The preliminary result came in on August 23, saying the skeleton was over 9,000 years old. The Army Corps of Engineers, custodians of the site, followed NAGPRA by contacting the relevant tribes (in this case, the Umatilla, Nez Percé, and Yakama; Chatters separately notified the Colville) on August 27.

And then the anthropologists held a press conference.

On August 28, the *Tri-City Herald* reported that while "initial indications were that the skeleton may ha[ve] been of European descent," it was obviously very old and indigenous. There was no indication of the bizarre racial furor the scientists would soon set off. But the story ended by quoting Jerry Meninick of the Yakama tribe compellingly about the sacred nature of the remains: "If Christians found a skeleton of someone who lived during the biblical book of Genesis, those remains would be revered."

Another slant to the story was, however, beginning to emerge. The following day, the *Seattle Times* ran a story with a very different theme. "Ancient Bones, Old Disputes—Is Kennewick Skeleton 'Asian' or 'European'?" it asked. This story contrasted the "narrow, European-shaped skull" with the "more rounded, Mongolian-type" of skull. It added rather anachronistically, "Scientists determine race by mathematically comparing measurements on several different points of the skull." The model now involved two migrations of different-headed peoples, "an early one of people with more Caucasian features—who today extend from Europe to India—and a later one with rounder skulls typical of contemporary northern Asians."

Chatters then contacted Rika Kaestle, then a graduate student at UC–Davis, and asked her to study Kennewick Man's DNA.

The last days of August were fateful for Kennewick Man. Seeing that a potentially intellectually valuable specimen might be rendered unavailable for study under NAGPRA, a leading anthropologist at the Smithsonian Institution invited Chatters to bring the skeleton east, away from the local Indians and the Army Corps of Engineers, and offered "Dr. Chatters airline accommodations in order to afford . . . an opportunity to conduct an extensive evaluation of the skeleton," according to court documents.

On learning of these plans, the Army Corps recalled the bones from the anthropologists, pending a decision about whose bones they were and whether the bones should be given to one of the local tribes under the federal law mandating a return of Indian bones to their descendants. Chatters was instructed to lock the bones in the county sheriff's evidence locker and go on with his life. "When I heard this, I panicked," he later recalled. "I was the only one who'd recorded any information on it. There were all these things I should have done. I didn't even have photographs of the post-cranial skeleton. I thought, 'Am I going to be the last scientist to see these bones?'"

Chatters's panic over his failure to document the find adequately is what set off the chain of events that became the Kennewick Man controversy. He had already consulted another physical anthropologist about the bones, and they had come to identical conclusions: the skeleton was similar to that of a Caucasian male.

Of course, no Caucasians—that is to say, people from Europe or western Asia—are known to have been in the state of Washington thousands of years ago. But regardless of how interesting the skeleton might be, the anthropologists were about to lose it to a bunch of people who didn't care about science and who just wanted to subject it to some superstitious mumbo-jumbo and destroy it.

So the anthropologists devised a way to make Kennewick Man very important. They cast him as the centerpiece in a literally earth-shaking theory—namely, that there were Caucasians in the New World thousands of years ago. And the carbon-14 date implied something even more earth-shaking: that Caucasians had been there *first*.

And they clothed themselves in the mantle of science and began to speak for it.

Thus, when the Umatillas filed formally on September 9 for re-patriation of the Kennewick Man skeleton, they were greeted with an extraordinary challenge to the legality of their claim. Kennewick Man, argued the scientists, did not fall under NAGPRA, because he wasn't an Indian ancestor. He was a Caucasoid, and thus the Indian tribes had no right to claim him under the law.

"The most immediate requirement is satisfying the tribes' desires to honor their ancestors and see those remains re-interred," a spokes-man for the Army Corps of Engineers said, but the anthropologist responded that the bones "have more European-like features, not characteristics often associated with early American Indians."

According to one scientist, "the Caucasian-like people represented by the skeleton died out in a severe Western drought about 9,000 years ago and were replaced a few thousand years later by the ancestors of today's Indians." According to another, "the earlier group inter-married with later arrivals more similar in appearance to today's northern Asians." Yet another speculated that the New World was prehistorically colonized by populations "hopping from England to Iceland to Greenland to Canada," which might explain the "Cauca-sian appearance of the early skeletons and . . . similarities between European and American stone tools."

On September 22, the *Seattle Times* mentioned the existence of non-Caucasoid features observed in Kennewick Man's remains, but they quickly vanished from the discussion.

On Monday, September 30, the stakes were permanently raised when the *New York Times* ran a story under the title "Tribe Stops Study of Bones That Challenge History" and brought the developing conflict over Kennewick Man to national attention. But what conflict was there? Under the relevant federal law, the bones belonged to the Indians, and that was that. Or it should have been.

The conflict, such as it was, had been manufactured by the scien-tists. They wanted the skeleton but didn't have rights to it. Suddenly, the skeleton became crucial, because it "adds credence to theories that some early inhabitants of North America came from European stock." But what theories were those? Where had they come from? What had they been based on?

"The finding," according to the *Times,* "does not necessarily suggest that white people were in North America before Indians, [Chatters] said. Rather, he said, it points to the possibility of mixed ancestry for today's American Indians." *Whatever that might mean.* But the article's headline made it clear what the issue was: "A Law on Indian Culture Sequesters Bones That Might be Caucasian."

The issue was race. The racial history of the world. Which race the skeleton belonged to. And how that could be scientifically determined.

And the Indians' superstitions could not be allowed to stand in the way.

Of the scientists' superstitions.

Anti-NAGPRA scientists circled the wagons to defend Jim Chatters. On October 16, eight of them—Robson Bonnichsen from Oregon State, Loring Brace from Michigan, Dennis Stanford and Douglas Owsley from the Smithsonian, Richard Jantz from Tennessee, George Gill from Wyoming, Vance Haynes from Arizona, and Gentry Steele from Texas A&M—filed a claim in federal court against the Army Corps of Engineers' plan to repatriate Kennewick Man to the Indians. Their central issue was that Kennewick Man didn't really belong to the people who are claiming him. NAGPRA stipulates that the remains have to go to people who have a "reasonable connection with the materials"—in other words, a "shared group identity which can be reasonably traced historically or prehistorically." Kennewick Man was going to the local Indians because he was a native American and they were native Americans who lived in the area and thus had a strong group identity based on their connection to the land and its earlier inhabitants. That certainly seems to fulfill NAGPRA's requirement of a "cultural affiliation" of the materials and the people.

But "cultural affiliation" and "group identity" are not scientifically ascertainable. They are the constructions of social history, not organic or natural properties. The law is written to permit an association like that between Yemeni Jews and French Jews, who may have nothing special in common that is physically perceptible and little behaviorally, so that it is difficult to distinguish between their association and the association between the Harrison Ford Fan Club in Omaha and the one in Yokohama. The two fan clubs are affiliated, and the identities of both are rooted in their admiration for Harrison Ford. Is there an objective basis on which to decide that the bond between two groups of Jews is valid and that between two groups of Harrison Ford's fans is not?

NAGPRA leaves the decision as to whether the bond claimed between Indians and specific remains is real or spurious to a committee.

Meanwhile the anti-NAGPRA anthropologists began to spin new origin myths for Native Americans. Robson Bonnichsen of Oregon State had already been quoted in the *Seattle Times* for his pet theory that "some of the first Americans came by skin boat from Europe by hopping from England to Iceland to Greenland to Canada," which, the newspaper was quick to add, was entirely unsupported by any "anthropological evidence" whatsoever. But it could explain both "the Caucasian appearance of the early skeletons" and "similarities between European and American stone tools."

Dennis Stanford, head of the Smithsonian's anthropology division, elaborated in March 1997. Based on superficial resemblances between early stone tools in America (Clovis points) and older ones from France (Solutrean laurel-leaf blades), Stanford suggested that the Indians came from France. "Now this is really an off the wall kind of idea right now," he added, "but it's one that I don't think we should ignore."

And as much as it should have been, it couldn't be ignored after Douglas Preston wrote it up in the *New Yorker* in June 1997. The original inhabitants of the New World came from France with their Solutrean tools across the Atlantic, and the Asian ancestors of the Indians came thousands of years later. Their skulls and their tools showed it.

The anthropology community was dumbfounded. Most students of stone tools found little similarity between Solutrean and Clovis points. For one thing, they were differently shaped; for another, they were of different sizes. For a third, they were made quite differently. The only similarity appeared to be that they both came to a point. The French-Clovis tool connection had not been published in peer-reviewed scientific forums, it had simply been put out in the *New Yorker*.

In fact, inferring cultural contact on the basis of a superficial similarity of technologies has a long history in anthropology. In the nineteenth century, it served to deny the creative abilities of indigenous peoples. There were pyramids in Mexico and pyramids in Egypt; the Mexicans couldn't possibly have built them themselves; so they must have gotten the idea from the Egyptians. Or perhaps from Atlantis. (Later, space aliens would fill the role.)

But there is, in fact, no basis for thinking that the Central American pyramids were conceived and erected by anyone but the indigenous inhabitants of the region, and likewise for the Egyptian pyramids. Similarly, there is no reason to think that Great Zimbabwe was built by anyone other than indigenous Africans, although colonial Europeans struggled mightily to resist that conclusion.

An even more insidious undertone of the Smithsonian anti-NAGPRA hypothesis is the linkage of biologies and technologies. Not only were the anthropologists dubiously inferring European peoples, but they were also dubiously inferring European tools, and mapping them on to one another, thus denying aboriginal Americans both a biological and a cultural identity!

Meanwhile, the Army Corps of Engineers had ordered all DNA work to cease on Kennewick Man and placed a gag order on the progress of the work to date. Kennewick Man did not appear to be the property of science, and the people he appeared to belong to didn't want scientists working on him.

But the scientists did have a cast of the skull and wished to make it known that the Indians were impeding the march of science. One widely repeated story in the newspapers and magazines involved trying to justify the racial propaganda for Kennewick Man by speculating on who he looked like, based on his skull. According to the *New Yorker*, Jim Chatters was watching *Star Trek: The Next Generation* and had an epiphany about Patrick Stewart, the actor who played Captain Jean-Luc Picard: "My God, there he is! Kennewick Man!" Unsurprisingly, then, when Chatters collaborated with a local sculptor a few months later to lay some flesh on the old bones, the result was a dead ringer for the English actor, "as Chatters predicted months ago," said the local newspaper on February 10, 1998.

This would serve to preempt debate over the race of the skull, by presenting a Caucasoid face rather than "race-less" bones. But again, there was exceedingly little science at work; an impressionistic description of Kennewick Man as resembling Captain Picard became a self-fulfilling prophecy. Not only did the earliest American tools come from France, but so did their users' faces!

From the bones, however, there is no greater resemblance to Patrick Stewart than there is to Patrick Ewing. In fact, the jaw, cheekbones, and eye sockets probably resemble those of the basketball star at least as much as they do those of the *Star Trek* actor. It wasn't widely publicized, but one physical anthropologist consulted by Chatters prior to the reclamation of the bones was reported as noting that "the shape of the skeleton's upper eye sockets . . . resembled the shape common to Negroid people."

The decision by a small group of anthropologists to challenge Kennewick Man's repatriation on the basis of its racial affiliation was very unfortunate. From a diverse assortment of characteristics, Kennewick Man was assigned to a single racial category. This is the philosophical problem of essentialism, in which you become defined by one or a few key parts, ostensibly representing your "essence," and the rest of you, the motley reality, is ignored. The essentialism underlying both racial categories and the assessment of racial affinity on the basis of skull form belongs to an earlier era, which anthropologists recall regretfully. Today, the anthropology community en masse rejects as pseudoscientific the notion of race as a natural category, and as equally bogus the assignment of individuals to such transcendent categories on the basis of their skulls. What exist are features and populations; sometimes one can map them onto one another, and sometimes one cannot.

As they tried to explain themselves to the anthropology community, the anti-NAGPRA team ran in circles. "[I]t was the very question of racial identity that drove the work," Jim Chatters informed the academic community in the monthly *Anthropology Newsletter,* but a few months later he asserted that "the scientific issue in the Kennewick Man debate is not race but human biological history."

Kennewick Man is a tragedy for several reasons. Most importantly, science is by now supposed to have transcended the premise that it must go on at any cost, no matter who is hurt or what anybody thinks. A science without regard for people's welfares and feelings is a science that is difficult to justify. Since people's identities are bound up in history, and their histories and identities are very significant to them, science treads on thin ice when it tampers with people's notions of who and what they are. That, of course, is a problem the Darwinian theory of evolution has always had in modern society. But at least Darwinism is a European idea being imposed on European society—it's even harder, and perceived more poorly, to impose a European scientific idea on a *non*-European society.

The problem is compounded by the quality of the science supposedly overturning the quaint myths of the indigenous peoples. Darwin's, after all, was a lot more solidly grounded than the science behind Kennewick Man.

The idea of a European settlement of the New World is also not new. The very first person to teach a class in anthropology in the United States was Daniel Garrison Brinton of the University of Pennsylvania. In his 1890 book *Peoples and Races,* Brinton explicitly rejected the popular notion that Indians had come to America from Asia; he thought they had come from Europe. The modern version differs in taking Europeans as the primordial Americans and seeing the Indians as having only subsequently come from Asia, thus robbing Native Americans of their claim to priority.

In fact, though, the transition of skull form through time has been acknowledged by generations of anthropologists. Brinton himself derided the use of skull form as a racial characteristic before the field of anthropology even assumed its modern form. In the case of the New World, Harvard's Earnest Hooton analyzed the skulls excavated from the site of Pecos Pueblo in 1930 and classified them into a diverse framework of types. He identified, for example, "pseudo-Negroids" and "pseudo-Australoids"—skulls that sort of resembled Africans or

Australians, but "pseudo" because he knew they were not African or Australian, but Indian. Better yet, he remarked that "a further type was characterized by compressed or medium malars [cheekbones], very long faces and deep mandibles, straight and often high-bridged and narrow noses, high orbits, and orthognathism. I called this type 'Long-faced Europeans' because it seemed to me rather un-Indian, if one may employ such an expression."

Indeed, one may, if only because it reveals the essentialist model under which one labors. The oddity seems to be that Hooton could openly acknowledge the plurality of physical types to be found among prehistoric Native Americans without resorting to imaginary ancient racial migrations as an explanation of it; whereas the anthropologists studying Kennewick Man seventy years later could *only* resort to racial migrations to explain him.

And Hooton's collection of Pecos Pueblo skulls was itself repatriated in the summer of 1999.

GIVE UP KENNEWICK MAN

I see three reasons to give Kennewick Man to the Indians and for the scientific community to forget about him.

First, he belongs to the Indians, not to science. That is the law. It is up to them to decide whether they want scientists to study him. Perhaps "the law is a ass," but whining about it will probably not be as efficacious as working to rebuild some confidence between the Indian and scientific communities.

Second, the scientists had their chance and muffed it badly. Not only did they trample callously and cavalierly on Native American beliefs and history—that is to say, not only were the scientists bad at the humanistic end of things—but their pronouncements were exceedingly speculative and poorly conceived. The irresponsibility of challenging the priority of Native Americans in America on such flimsy evidence harks back to the authoritarian nonsense espoused by geneticists of the 1920s. It was never an issue of Indian beliefs versus science, but of Indian beliefs versus *pseudo*science. Perhaps the next iteration of scientific thinking and writing about Kennewick Man and the peopling of the Americas will be deeper and more conscientious, but like young children, scientists will have to earn their second chance.

And third, Kennewick Man simply isn't that important. One specimen is never that important. If it is a unique and irreplaceable specimen, then it is scientifically valueless, for that would make it a singularity of nature, in layman's terms, miraculous. And science doesn't deal in miracles, it deals in regularities. There will be another Kennewick (as indeed there already are a handful of similar remains), and if science repairs its community relations after the disgrace of Kennewick Man, then perhaps it will have a chance to study the next skull in detail. The irreplaceable value of Kennewick Man lies in its value for the careers of the scientists, not in its value for the progress of science.

What binds this into the theme of molecular anthropology is the conception of genetic descent being invoked all throughout the argument, particularly by the scientists. The scientists argued that the Indians claiming him were not his descendants because they were of different races. That claim is nonsensical; but it could be rendered sensible by saying simply that Kennewick Man was "physically different" from living Indians. That statement might be true, but it would be trivial, for there are several reasons why that might be the case—and invoking major invasions of completely geographically distinct populations is only one of them.

But the fact is, at this level, all discussion of descent is metaphorical. We do not, and cannot, know whether Kennewick Man had any descendants *literally.* (Although the spear point in his pelvis would certainly have made procreation painful!) The issue is the *group* he belonged to, or that we assign him to, and its relationship to today's groups. Scientific or genetic ideas of identity and descent are not the heart of the matter, *for descent is not being taken literally here.* Thus, any claim of ancestry for Kennewick Man is in the realm of folk heredity, in the cultural ideas of relationship among constructed groups. It isn't that Kennewick Man is excluded from Umatilla ancestry because he was Caucasoid; or because they probably have not lived in the area continuously for 9,000 years; or because his people died out and were replaced by other physically different groups of indigenous Americans. It's that *all claims of ancestry and descent from Kennewick Man are nonliteral and nonscientific; they are metaphorical.* And we have no objective, scientific basis on which to judge one metaphor's validity as against another's. The question of descent from Kennewick Man thus falls outside the domain of science.

Kennewick Man has different significance for the two groups that want his remains, and his importance as a symbol to Native Americans, I would argue, outweighs his importance to the scientists as a basis for thoughtless and irresponsible speculation. It has never been good for science to trample callously on people's views of themselves and their place in the universe. All that does is make people angry, resentful, and disoriented. And science was always supposed to improve people's lives, not make them worse.

Kennewick Man lay at the crossroads of the sciences and the humanities. He represented a confrontation between the politics of identity and human rights, on the one hand, and an archaic and transgressive science, on the other.

CODA

In February and March of 1999, a blue-ribbon panel of scientific experts was charged by a U.S. District Court to examine Kennewick Man and to decide whether the remains actually fell under NAGPRA. This was in itself a victory for the scientists, although the specific charge was specifically to perform nondestructive work, a concession to Indian sensibilities.

The panel's conclusions were predictable. It began by acknowledging that "the geographic groupings or races seen among modern peoples are at best fuzzy and at worst non-existent when examining [ancient] populations world-wide." Kennewick Man looks different from modern Indians, but isn't Caucasoid by any stretch. He has some cranial features commonly found in Europeans, others commonly found in east Asians, and still others indeed found in Native Americans. Most charitably, the panel determined that he had some "white" features, which are actually also found quite broadly, it noted, among other peoples, notably Polynesians. A statistical analysis found him most closely resembling Polynesians and southern Asians.

The panel was too kind to identify the racialized science surrounding Kennewick Man for the bunkum it was—an appropriately measured tone for a blue-ribbon panel report. It did not say that the Americas had been peopled by Polynesians—all the similarity means is that Polynesians have a more generalized cranial form, and fewer specialized idiosyncracies of the skull, than other comparative groups. But if Kennewick Man is not Caucasian, there's no longer anything to preclude him from being an Indian ancestor—which means that he ought to fall under NAGPRA. After all, as the panel was told by the government, "the term 'Native American' is clearly intended by NAGPRA to encompass all tribes, peoples, and cultures that were residents of the lands comprising the United States prior to historically documented European exploration of these lands."

About the same time as the Kennewick Man report came out, word came from Brazil that anthropologists there had the skeleton of an ancient woman they called "Luzia" after the 3-million-year-old australopithecine "Lucy." Luzia was three million years younger than her namesake and belonged to a different species, and her story was modeled not on that of Lucy but on Kennewick Man's.

Luzia, reported the London *Times* on August 22, 1999, is 12,000 years old and doesn't look at all "Mongoloid." In fact, said the scientists working on the skull, Luzia's features "are similar to modern-day [Australian] Aborigines and Africans and show no similarities at all with Mongoloids from east Asia and modern-day Indians." And they purported to find links between the material culture of Australians and the inhabitants of Tierra del Fuego!

Thus the lead sentence from the London *Times* article: "The first people to inhabit America were Australian Aborigines—not American Indians."

The skulls change, but the story remains the same.

Frankly, I find the use of science to undermine indigenous people's sense of identity crass and boorish. It represents the worst of what science has to offer—a threatening, insensate, and destructive ideology.

Eleven

IS BLOOD REALLY SO DAMN THICK?

WHILE KENNEWICK MAN'S skeleton may seem to be a subject remote from the 98% genetic equivalence of humans and apes, it really isn't. It's all fundamentally about descent and relatedness—how I identify my relatives and what their relationship to me means. In many cultures, this is usually expressed through the medium of blood.

Blood is one of the most powerful substances the human mind has ever invented.

I don't mean that, of course, literally. People didn't invent blood. It existed as a biological substance before people did.

But as such it has no meaning. People gave it meaning. It became the emblem of a warrior's prowess, the wondrous life force of a god, the taboo monthly flow of a woman, and the link between the generations. It became the Eucharist of Jesus, the hallucination of Lady Macbeth, and lunch for Dracula.

That is what we mean when we say that blood is a "construction." It has importance or significance that is not a direct consequence of its biological nature—of the mixture of plasma, erythrocytes, basophils, neutrophils, lymphocytes, and the rest.

Blood, rather, is a highly sacred substance.

If that doesn't sound scientific, that is precisely the point. Blood is culturally constructed as a sacred substance, and its significance stems from the meaning it contains in the human world of symbols and semiotics—not in the material world of cells, molecules, and biology.

We saw a similar situation with respect to race: the fact that the natural patterns of biological or genetic variation do not permit us to subdivide the human species into a small number of reasonably homogeneous groups has little to do with the importance of race in modern society. Race affects your income, lifespan, and health and suffuses every aspect of American life—and yet it has no biological reality. We cling to the paradox of blood (part natural, part construct) as we do to the paradox of race (geographical variation is real, but essentialized human subspecies aren't) and the paradox of our place in nature (part ape, part angel, in spite of the fact that apes exist and angels don't). What makes them all paradoxes is that both parts are important, regardless of their ontological status.

The most widespread meaning of blood is as a substitute for heredity—the "blood" in "of royal blood" or "blood relatives." The word is also commonly used racially, as in "Negro blood." In the classic musical *Show Boat,* based on the Edna Ferber novel, the theatrical troupe find their star couple accused of miscegenation. Learning of the accusation ahead of time, the husband cuts his wife's hand with a penknife, and licks her bloody palm in full view of the troupe. A few minutes later, when the sheriff makes the formal accusation of miscegenation, the husband responds that although he looks white, he has Negro blood in him—and of course all the sympathetic witnesses swear to the literal truth of the statement.

The sheriff, of course, was using blood as a metaphor for heredity, in the folk sense we discussed earlier. "One drop of Negro blood makes you a Negro in Mississippi," he intones. (But who didn't know that?) And the couple are spared for the moment because they respond literally.

Relatives share blood, and racial relatives, although distant, share traces of it, under this model. And this model is essentialist because it holds that no matter how infinitesimal the dilution, there will always be a detectable element—an essence—that reveals the "other-ness" in an individual's makeup.

The great contribution of the Mendelians, however, was to demonstrate that heredity is not essentialist but probabilistic. Every genetic element has a fifty-fifty chance of being passed on—no more, no less.

Let's say Mom and Dad have a baby girl, who for some reason later either wants to, or has to, prove her racial identity. The ultimate arbiter of heredity is science, the science of genetics. Mom and Dad look racially appropriate, as does the girl, but for some reason that's not enough. So the girl is subjected to a genetic test to look for a specific genetic marker that her race has and that others don't.

If we take her family back two more generations, we know she has four grandparents and eight great-grandparents. Did all eight have that genetic marker?

Because if one didn't, there is a small but significant chance that the girl is also missing that crucial genetic marker; and it would genetically identify her as racially "wrong."

And that's precisely the problem. If one great-grandma was different, whether because she was from somewhere else or because the generalization was incorrect to start with, and she inherited the genetic marker from one of her ancestors, then it has a fifty-fifty chance of being passed on in each generation.

In other words, you could have a child who looks just like you and your spouse, but fails a genetic test on account of one great-grandparent (who may have been just as much of the same "race" as the other seven). That's why you can't do a genetic test for race, because race and genetics don't map onto each other perfectly. Race is an essentialized property, and genetics is probabilistic.

Thus, when the Vermont House of Representatives began to consider Bill 809, drafted by Republican Fred Maslack in late 1999, to "establish standards and procedures for DNA-HLA testing to determine the identity of an individual as a Native American, . . . [and to provide] conclusive proof of the Native American ancestry of the individual," it was greeted with astonishment by both Indians and geneticists. First, because it removes from the Indian nations the right to define their membership, and makes it a state of nature accessible instead by the U.S. government; and, second, because it is scientifically impossible.

There is not, and cannot be, a single genetic marker that, say, all Africans have and no non-Africans have. Because of their biological propensities for interbreeding, and their histories of having done so, all human groups are culturally bounded. Even a term as natural-sounding as "Africans" unites physically and genetically diverse peoples and is usually taken to include people who themselves have never been to Africa, but whose ancestors came from there, and to exclude people whose families have lived there for hundreds of years. Thus, one can hear surprisingly casual talk about white South Africans as non-African, by people who would never consider white Americans as non-American.

But that same confusion doesn't exist genetically between chimpanzee and human. All humans have a pair of large chromosomes (#2) that no chimpanzee has. It is a correlate, not a cause, of humanness, but it permits an unambiguous diagnosis of the allocation and ancestry of the cell in question.

ESSENTIALIZING BLOOD

Essentialism is such an ingrained cultural value that it surfaces surprisingly and unfortunately often in population genetics, where it is paradoxically most fundamentally undermined.

A widely publicized study in early 1997 tracked the Y chromosomes of Jews to examine the hereditary priesthood that most people don't even know Jews have. According to the Bible, Moses's brother Aaron became the progenitor of a lineage, the Cohanim, entrusted with special duties in the Temple in Jerusalem. With the destruction of the Second Temple by the Romans in the first century, the formal functions of the priestly lineage came to an end, although some duties and privileges remain in local synagogues for people with the surname Cohen or its cognates.

A group of geneticists—I'm not sure whether more pious than credulous, or vice versa—examined the configuration of two genes on the Y chromosome, which is passed on from father to son, like the priesthood. And they found that of Jews who answered "yes" to the question of whether they are priestly, 54% had one particular configuration; while of Jews who answered "no," only 33% had this configuration. In both groups, it was the most common configuration.

The obvious interpretation, from the standpoint of anthropology, might be that the Cohanim simply represent a somewhat more inbred group than the non-Cohanim, because many of them have the same last name, which means they are more closely related to one another than random Jews would be.

These scientists, however, took a different tack. They figured that, taking the biblical story of the origin of the priestly lineage at face value, all male descendants of Aaron would have Aaron's Y chromosome. That Y chromosome would today be the most common one in the Cohanim, implying that the 46% of Cohanim with a different Y chromosome are nothing but pretenders. The study's first author said bluntly to the *New York Times,* "The simplest, most straightforward explanation is that these men have the Y chromosome of Aaron."

I am reminded of a line of H. L. Mencken's: "To every complex problem, there is a simple solution . . . and it's wrong."

In this case, we certainly can't prove their interpretation to be wrong. But of course, it's not our burden in science to do that. It's their burden to prove that they have done the adequate controls and given the most reasonable interpretation. And in science, an extraordinary claim (like proving the Bible true!) requires extraordinary standards.

A reasonably sophisticated audience, on the other hand, appreciates the Exodus narrative as an origin myth of the Jews, and this part specifically as a politically value-laden founding myth of the priestly caste. Not only are the priests thereby justified back to the Exodus by virtue of the narrative, but in accepting it, the geneticists are also piously justifying their religious beliefs. Aaron is, after all, reputedly the brother of Moses—he of the bulrushes, the burning bush, and the stone tablets. Aaron's Y chromosome is also that of the Lawgiver.

In most scholarly contexts at the turn of the twenty-first century, one might be quite stunned to see a scientific study that begins by assuming the literal truth of the Bible. I can imagine with some trepidation an historical linguistic analysis assuming the Tower of Babel (Gen. 11:9); a study of negative mass assuming the ascension of Enoch (Gen. 5:24); a study of primate biogeography assuming Noah's ark (Gen. 8:19); or a study of cellular hypoxia assuming the resurrection of the body of Christ (Luke 24:15).

That people with similar surnames tend to be genetically more closely related to one another than to people selected at random is a well-known principle in biological anthropology. It is called "isonymy" and can even be used as a noninvasive surrogate measure of the degree of inbreeding in a population. For example, a famous survey of the Pennsylvania Amish in the 1950s, where a rare genetic disease called Ellis–van Creveld Syndrome was surprisingly prevalent, found that of 1100 ostensibly unrelated families, 23% were named Stoltzfus, 12% were named King, 12% were named Fisher, and so on—six surnames accounted for 72% of the families. The knowledge, therefore, that the sample of Cohanim contains a few surnames considerably overrepresented should suggest that this genetic result might be tracking isonymy, and therefore inbreeding, and not necessarily revealing the constitution of the single mythic founder of the priestly lineage.

Consequently, the second study by the same geneticists, utilizing twelve markers on the Y chromosome, and finding a most common form shared by 51% of self-identified Cohanim and only 12% of other Jews, tells us exactly the same thing. They have proven robustly that that the Cohanim are somewhat distinct from, and somewhat more inbred than, other Jews. That is, the Cohanim are more genetically similar to one another than to other people.

In fact, the pattern detected among the Y chromosomes of the Cohanim is precisely the pattern we would expect from a group of people chosen largely on the basis of the overrepresentation of a few surnames. It is the same pattern you would expect of a sample in which half the people were named Horowitz or Bernstein. (In fact, that is precisely the pattern detected in a recent study of people named Sykes in three rural English counties.) In the absence of controls involving samples comparably overrepresenting Horowitzes or Bernsteins, we have no way to know just what the Cohanim pattern does represent. And in the absence of comparative knowledge of the frequency of these arrays in non-Jews, the information is decontextualized and largely meaningless.

It is thus certainly gratuitous to assume, along with the reverent geneticists, that the biblical narrative of the founding of the priesthood is literally true and has now been validated genetically. It is conceivable alternatively that the Cohanim are of heterogeneous origin; that they comprise a population and not a single genetic form with a lot of pretenders; that the genetic similarity within the group may be a simple function of recent historical relatedness in the construction of the sample; and that there might be no single primordial priestly configuration, even though this methodology combined with essentialist assumptions will invariably identify one.

More important, the authors of the report find themselves in the middle of an identity controversy, as people want to know authoritatively if they are "really" Hebrew priests or not. Well, of course, nobody's a Hebrew priest; there hasn't been a priesthood for centuries. Nevertheless, in spite of how shaky the inference is, these genetic data are culturally invested with religious authority.

The construction of identity is a political arena in which genetic data should be regarded with considerable caution! Perhaps the most bizarre aspect of this particular story is that the *New York Times* reported further on May 9, 1999, that the "Cohanim" Y chromosome is also found among the Lemba, a South African group with an origin myth associating them with the Jews.

And therefore, they are genetic crypto-Jews.

Perhaps so, but since the story and the data had not yet been published, reviewed, submitted, or even written up for a scientific audience, perhaps it would be wise to suspend judgment. Those who are forced to trumpet their research in the press, rather than in a scientific forum, usually have a good reason for doing so. (Such as, they can't really prove their point because they haven't done the appropriate controls or ancillary studies.) In this case, it would be nice to know how prevalent the "Cohanim" Y chromosome is among

Middle Eastern peoples generally, particularly among the descendants of the ancient seafaring nations, which undoubtedly affected the gene pools of their client states.

Among them probably the Lemba.

Indeed, when the work finally did appear in a scholarly forum (the *American Journal of Human Genetics*), the authors were obliged to concede that generalized gene flow from the Middle East—not necessarily from Jews—was what they had found.

WHAT IS A RELATIVE?

Clearly, then, among the most mystified concepts in the modern world is that of relatedness. I know my relatives, and I love them, because they stand in a special relationship to me.

We share blood.

Of course we don't share blood *literally.* But we do share something, don't we? What is it that relatives share that nonrelatives don't?

That is, in fact, a surprisingly subversive question. The social anthropologist David Schneider observed that Americans believe that relatives share some manner of "biogenetic substance" with each other, but left the exact nature of that substance unexplored. He simply presented it as a cultural fact, that kinship balances the relations of "biogenetic substance" alongside relations of law, or marriage. He referred to it as the axiom that Blood Is Thicker Than Water.

But it is a question that a molecular anthropology can address. Is there a biogenetic substance in any literal, natural sense that I share more of with my sister than with my cousin? Everyone knows that "blood" is just a metaphor; it's not as though relatives regularly get transfusions from one another.

Perhaps the biogenetic substance is chromosomes? Actually, however, the basic genetics of crossing-over dictates that segments of chromosomes will be scrambled every generation. Of the twenty-three chromosomes I pass along, any particular chromosome my daughter has inherited from me is quite different from the one I inherited myself; it is in fact a mixture of bits and pieces from the two chromosomes I inherited from my parents.

And the biogenetic substance cannot be specific segments of chromosomes, because there is no guarantee that any particular segment of my mother's chromosome is present in my daughter. The particular chromosome segment of interest that my mother passed on to me only has a fifty-fifty chance of being passed on to my daughter.

That leaves us with the well-known sociobiological calculus of shared proportions of genes—50% with a sibling or parent or child; 25% with a niece or nephew, and a subset of uncles and aunts; and 12½% with a cousin. (Sociobiologists invoke this to explain altruism.) But the concept of shared proportions of genes is metaphoric as well—that's rather *less* well known—because, of course, we share 98% of our *literal* biogenetic substance with chimpanzees and gorillas.

This shared-proportions-of-genes business is derived from theoretical population genetics and didn't originally involve the sharing of a percentage of substance—as it is too commonly articulated. Rather, however, it represents the *probability* of two people inheriting the same specific chunk of heredity—the same gene, or DNA segment—from the same common ancestor. In other words, it is a calculation of the probability of "identity by descent"—that is the phrase from population genetics—for any particular gene. It would also represent the average proportion of genes in two people that are identical by descent—but that proportion would necessarily comprise a ridiculously small complement of the genes, the DNA, that they actually *do* share, given the 98% shared with apes. Most DNA you have is, of course, identical to anyone else's.

How this calculation managed to become enmeshed in a discourse of overall proportion of shared biogenetic substance, and from there, to become the bedrock of kin selection in sociobiology, is a good question for a social historian of science. The point is, however, that genetically we do not share "a proportion of genes" in any literal sense with specific relatives; rather, there is a probability that we both possess the identical gene, inherited independently from the same ancestor.

What, then, do relatives hold in common?

The only attributes that relatives share—the literal nature of this biogenetic substance—are revealed in precisely the manner in which the great mathematical geneticist Sewall Wright originally formalized the genetic issues. And that is, the biogenetic substance can only be the common ancestors themselves.

That is what relatives literally share. Ancestors.

Ultimately, as with Santa Claus, there is no biogenetic substance. The concept is a survival, a folk ideology of heredity. There's DNA and there are probabilities of sharing some, but no tangible genetic stuff divisible among kin and distinguishing them or bounding them from non-kin. There is no genetic test for kinship.

Kinship is not a genetic property.

As anthropologists have known all along, kinship is constructed, for we make very arbitrary decisions about things as simple as where to stop calculating. While we regard a sister as a relative, a second cousin is a relative outside the bounds of incest, and it's not clear that a fifteenth cousin is really a relative at all. All three of them might share the same genetic marker, or they might not. What they share are merely different probabilities of having that marker.

With or without the Cohanim Y chromosome, the South African Lemba might or might not be related to the Jews.

So in a literal, natural sense, relatedness is a just a mathematical abstraction. It has no real, bracketable biological properties. In a cultural, meaningful sense, kinship is a way of defining social networks, establishing obligations, and organizing the transmission of property across generations. But it is not based on the shared possession of particular physical or genetic attributes.

A LAST LOOK AT KENNEWICK MAN

What, then, are we to make of the suggestion that Kennewick Man's DNA should be studied in order to determine who he was and who owns him?

In late 1999, a blue-ribbon panel reported back to the U.S. Department of the Interior that since Kennewick Man appeared to be 9,000 years old (based on a replication of the radiocarbon date), he had to be most reasonably regarded as Native American, and therefore subject to NAGPRA. The last gasp of the scientists who wanted to study him was to clamor for DNA tests—the kinds of studies that Rika Kaestle had begun in good faith but had not been allowed to complete.

But why? Why demand DNA tests at this point, against the wishes of the Indians?

The first question is whether you believe that nothing, like sacred Indian beliefs, should stand in the way of science. But that question has already been answered—there are things that can legitimately impede the progress of science, things like human rights. Just because something might be interesting to learn (like just how much radiation it takes to sterilize the ovaries of a Jew), doesn't mean that you can simply carry out the experiment (as Josef Mengele did at Auschwitz, to his lasting infamy).

The question then is to establish what the possible benefits are for scientific knowledge, and to weigh them against the possible harm. The harm is the desecration of remains Amerindians regard as those of an ancestor, the continuing irresponsible assault on their identities as indigenous original inhabitants of North America, and the resulting alienation of Native communities vis-à-vis the anthropological community and federal agencies.

And the benefit to science of grinding up bits of Kennewick Man and doing a DNA test?

We'll come back to that.

Whatever the questions that Kennewick Man's DNA could resolve, let's say you decide that trying to solve them outweighs the possible harm. So you desecrate the remains and grind up some of the bones. What is the most likely outcome?

Quite simply, nothing. DNA can be extracted from old bones a thousand years old with a reasonable degree of success (about 70%), but Kennewick Man is nine thousand years old. The chances of actually recovering any DNA from him are rather small. So the most likely outcome, even if you decide to override the concerns of the Indians, is that you will have destroyed the material in vain.

Now, let's say you're lucky and skillful enough to get some usable DNA. It would have to be mitochondrial DNA, which exists in many copies per cell and evolves rapidly. What can you do with it?

Well, there are only two questions you could conceivably ask with Kennewick Man's DNA. First, is it Native American DNA? It turns out that five clusters of mutations in the mtDNA of Native Americans tend to define populations of the New World. I say "tend to define" because those five clusters are neither exclusive to Native Americans nor 100% characteristic of Native Americans. So finding that Kennewick Man has one of the five mtDNA clusters doesn't actually tell you he's Native American, and *not* finding it doesn't tell you he *isn't* Native American. In any event, that question has already been decided, so this seems hardly worth the effort.

On the other hand, doing a DNA test, argue some of its proponents, might allow you to tell what tribe he belongs to and therefore who should receive his remains.

On the other hand, if the tribes in question don't allow you to do DNA tests on them, then you can't possibly say anything about Kennewick Man's affiliation to them. More the point, though, is that the tribes themselves are historically and politically constituted entities. Their boundaries are porous and their durations ephemeral over the scope of human history and prehistory. Consequently, closely related tribes don't differ genetically from each other in any but the most exceedingly subtle fashions. To try and allocate Kennewick Man to one of them on the basis of his DNA would be a fool's errand.

So why not just let him rest in peace?

But they wouldn't. Federal authorities decided in early 2000 to go ahead with genetic tests on Kennewick Man in spite of the likelihood of finding nothing interesting and desecrating a sacred relic to accomplish it. Rika Kaestle would get her chance to study Kennewick Man's DNA after all; now the government was asking her to do it.

And when the smoke cleared in the autumn of 2000, they didn't find any DNA after all, and Secretary of the Interior Bruce Babbitt finally recommended that Kennewick Man be given to the Indians. Those scientists are still protesting.

Species, on the other hand, are not constructed entities: they are natural lineages bounded in time and space by participation in a common gene pool. Humans constitute one, chimpanzees another. And they appear to have had separate biological histories for about seven million years.

That recognition is the culmination of Darwin's revolution, the acknowledgment that species have histories, and their patterns of resemblance indicate aspects of their relatedness to one another. That, in turn, was the culmination of an earlier revolution, the acknowledgment that observation and experimentation—the empirical core of science—constituted a powerful manner of understanding the universe. Such an understanding would comprise a more fundamental reality than had ever been previously apprehended; and that apprehension was fundamentally good, because it would afford a means of improving the quality of life for all, by the application of this knowledge of reality via technology. And moreover, it is therefore desirable to promote science and scientific understandings as widely as possible.

This was the mandate promoted in the sixteenth and seventeenth centuries by the likes of Francis Bacon in response to the defensive position of the clergy and the conservative position of the common folk. Science as a way to new knowledge was appropriately perceived as threatening by the custodians of the old ways of knowledge—revelation and fiat.

But now we know that living and extinct species are genealogically connected, and that our history is intimately bound up with that of the apes.

I've never quite understood what is so threatening about that, in and of itself. Aside from the fact that it seems to go against biblical narrative—but so do a lot of things. After all, we know that English and Spanish weren't spoken at the foot of the Tower of Babel—they evolved in historical times from Germanic and Latin, respectively. We know that Jesus didn't survive Herod the Great's order to massacre the innocents (as St. Matthew says) if he was born in Bethlehem during the census of Quirinius (as St. Luke says), because the census occurred in 6 A.D., when Herod the Great was already ten years in his grave (as historical records say).

So what's threatening to religion about chimpanzees?

As with other forms of kinship, the common ancestry of humans and apes constitutes an historical and social narrative, a story about where you came from and ultimately about what and who you are. Like other kinship narratives, it locates you within a social universe of others, and identifies a certain class as meaningfully similar to you and the rest as meaningfully dissimilar from you.

Quite a responsibility for science to assume. And it is the core of the paradox of the ongoing creation-evolution war: science doesn't care about the risk of alienating people from the belief systems that orient them in life. It cares simply to describe "what is"—the descent of our species from an ape stock, and our intimate kinship to living apes.

People, on the other hand, do care about their orientations. Learning, for example, that you were adopted, and that your parents aren't your biological parents, and that you don't have the nexus of kinship to your relatives that you thought you had, would be highly disorienting. The scientific origin narrative is simply the opposite, from which we learn that we indeed have parents and relatives, and do not stand isolated from rest of the species on earth.

But it is no less disorienting.

Consequently, I think that creationists are entitled to a degree of sympathy they don't often encounter in the scientific community.

UNCONVENTIONAL THOUGHTS ON CREATIONISM

Biologists have a reason to get defensive about creationists, who do seek to undermine science education in America. They successfully prosecuted John T. Scopes in Tennessee in 1925 ("it shall be unlawful for any teacher in any of the Universities, Normals and all other public schools of the State which are supported in whole or in part by the public school funds of the State, to teach any theory that denies the story of the Divine Creation of man as taught in the Bible, and to teach instead that man has descended from a lower order of animals"), and although he successfully appealed and got off on a technicality in a higher court, they made their point. Evolution was a distasteful philosophy that went counter to the teachings of the Bible and needed to be fought.

And for the next thirty-five years, their point was acknowledged in high school biology textbooks, which generally ignored the subject of evolution. It wasn't until the shock of the Soviet satellite Sputnik stimulated a heavy investment in science education that evolution made a big comeback in the curriculum.

By the mid-1970s, fundamentalist Christians had a series of loosely affiliated organizations in place, most notably the California-based Institute for Creation Research, which actively sponsored undermining the teaching of the theory of evolution. Local university chapters of fundamentalist student groups freely distributed literature and were able to choose from a stable of speakers. Prominent among these was the diminutive biochemist Duane Gish, author of a series of polemics against evolution and a very competent orator, who toured college campuses with a slide show, occasionally debating (and usually trouncing) his opponents. Gish would show a picture of a half-whale, half-cow and demand to know where this "transitional form" could be found; he'd talk about the fraudulent fossil Piltdown Man, and about Nebraska Man, an obscure, transient misinterpretation of a peccary tooth for an early human tooth. And he would invoke bizarre and seemingly random data to prove his point—footprints of humans and dinosaurs in a Texas riverbed; the atomic structure of the element polonium; the anti-predation defenses of the bombardier beetle.

The grass-roots campaign to call evolution into question was so successful that by the late 1970s, Arkansas passed a bill mandating "equal time" for evolution and creationism. Creationist logic was inconsistent here—sometimes they should be given equal time because both were religious, and sometimes they should be given equal time because both were scientific. Nevertheless, the law was overturned as unconstitutional in federal court in 1978, as Judge William R. Overton concluded that evolution was science and creationism was religion.

The 98% genetic correspondence of humans and chimpanzees does have a consequence with which hard-core creationists must wrestle—namely, that either humans and chimps do share a recent common ancestry, or else they have been independently zapped into existence by Someone lacking a great deal of imagination.

At the turn of the millennium, creationists are a diverse lot. Some believe in a young earth—they reject not only genetics, biology, and anthropology, but geology and astrophysics as well. To them the universe is only a few thousand years old, as the chronology of Genesis implies, a position even William Jennings Bryan stopped short of at the Scopes trial. (He articulated the position freely and wasn't tricked into it, as the play *Inherit the Wind* incorrectly suggests.) Others are, like Bryan, "old earth creationists" and implicitly accept geology and astrophysics, but reject biology. And still others accept the ancient cosmos and much of biological evolution, but refuse to accept the possibility of continuity between very different kinds of creatures—most especially humans.

One of the most prominent exponents of the last form of creationism is an affable law professor from Berkeley named Phillip Johnson. Johnson has little argument with the possibility that "a population of birds happens to migrate to an isolated island" as a result of which "a combination of inbreeding, mutation and natural selection may cause this isolated population to develop different characteristics from those possessed by the ancestral population on the mainland." To Johnson, this is so mundane as to be "uncontroversial"—but for a significant reason: it "has no important philosophical or theological implications."

Johnson calls attention to the casualness with which evolutionary science dispatches meaning from life and the superficiality with which scientists attempt to fill the intellectual vacuum they thereby create. The central problem is a methodological assumption on the part of science that he calls "philosophical naturalism"—namely, that the natural world must be explained in terms of itself, and not in terms of occult forces or supernatural beings.

Why is this assumption necessary? Because that is simply how science works. It tries to make sense of the natural world, and if the forces operating in it are supernatural or occult, they are by that very fact unknowable and capricious. If such forces operate, they defeat the purpose of science.

Which isn't to say they don't exist, only that we can't do science with them, or on them. So we ignore them.

And science does seem to do well with that assumption. A large class of diseases do seem to be caused by germs and not by evil spirits. And they respond better to antibiotics than they do to prayer.

On the other hand, placebos also work a significant percentage of the time. And we don't know precisely why. We haven't explained everything.

Which brings us back to evolution.

William Jennings Bryan, grilled by Clarence Darrow at the Scopes trial, is not represented well by the character called "Matthew Harrison Brady" in *Inherit the Wind*. For one thing, the exchanges were often more interesting, at times resembling bits of an Abbott-and-Costello routine, than they are as abbreviated or invented on the stage and screen. For example: the real Darrow tries to show off Bryan's ignorance of other religions, and asks him about Buddhism. Bryan responds:

> BRYAN: Buddhism is an agnostic religion.
>
> DARROW: To what? What do you mean by "agnostic"?
>
> BRYAN: I don't know.
>
> DARROW: You don't know what you mean?
>
> BRYAN: That is what "agnosticism" is—"I don't know."

The crux of the matter, however, was Bryan's rejection of scientific expertise in his understanding of deep time. Darrow asked if he knew of any civilizations older than 5,000 years. Quite poignantly, Bryan responded.

> Well, so far as I know, but when the scientists differ from twenty-four millions to three hundred millions in their opinions as to

how long ago life came here, I want them to be nearer, to come nearer together, before they demand of me to give up my belief in the Bible.

In other words, if my Bible is more important to me than the age of civilization is, why hassle me about it? And moreover, what is the basis for asserting the authority of science over anything else when science is itself internally inconsistent?

Bryan's politics have been obscured by his anti-science legacy, but they were surprisingly modern. He was a radical anti-imperialist in a time of rife exploitation and empire-building and a staunch pacifist in an era of American machismo vis-à-vis Mexico and the Philippines. He ultimately resigned as secretary of state when it became clear that the United States would be unable to stay out of World War I.

His consistent political stance was in fact concordant with his stance on science. Very significantly, Bryan lived in the heyday of the scientific-political hodgepodge retrospectively labeled "social Darwinism," when scholars commonly invoked biology and Darwin in support of war and imperialism, the two things Bryan opposed most strongly.

If our descent from the apes meant that science justified—indeed mandated!—the bellicose colonialism he opposed, it's no surprise he would come to oppose evolution as well.

In a speech in 1900, Bryan argued:

> There are degrees of proficiency in the art of self-government, but it is a reflection upon the Creator to say that he denied to any people the capacity for self-government. Once admit that some people are capable of self-government and that others are not and that the capable people have a right to seize upon and govern the incapable, and you make force—brute force—the only foundation of government and invite the reign of a despot. I am not willing to believe that an all-wise and an all-loving God created the Filipinos and then left them thousands of years helpless until the islands attracted the attention of European nations.

But consider the literature representing evolution to the public. The first generation of evolutionary anthropologists presented civilization as a "stage" beyond savagery and barbarism, where other societies in other parts of the world still languished. Evolutionary considerations dictated that since they had not risen to the point of being able to govern themselves, they were most appropriate as colonies or corpses. A distinguished British paleontologist named William J. Sollas, in his 1911 book *Ancient Hunters*, interpreted human prehistory as a series of invasions and conquests: "The vanished Paleolithic hunters have succeeded one another over Europe in the order of intelligence: each has yielded in turn to a more highly developed and more highly gifted form of man." Remnants of these earlier peoples, he adds, are to be found as Australians, southern Africans, Eskimos, and native Americas. And in case the political import is still unclear to his readers, he goes on:

> Justice belongs to the strong, and has been meted out to each race according to its strength; each has received as much justice as it deserved. . . . It is not priority of occupation, but the power to utilize, which establishes a claim to the land. Hence it is a duty which every race owes to itself, and to the human family as well, to cultivate by every possible means its own strength . . . [lest it incur] a penalty which Natural Selection, the stern but beneficent tyrant of the organic world, will assuredly exact, and that speedily, to the full.

Sollas was far from alone in his political interpretations of evolution. A distinguished Canadian paleontologist, William Diller Matthew, developed general principles of biogeography in which Europe and western Asia were the cradle for more highly evolved types of organisms—including people—which then go out and colonize other areas. The Darwinian Karl Pearson argued that "a capable and stalwart race of white men should replace a dark-skinned tribe which can neither utilize its land for the full benefit of mankind, nor contribute its quota to the common stock of human knowledge."

Bryan was aware of this literature. When he read Darwin's *The Descent of Man* in 1905, he commented insightfully that the material served to "weaken the cause of democracy and strengthen class pride and the power of wealth." For Bryan the last straw came in 1917, when he read a book by Vernon Kellogg, an American biologist, which warned specifically of the manner in which the Germans were invoking Darwinism in their war effort.

Phillip Johnson's modern antagonism to evolution is rooted in the same very real problem. It is a question of the relationship between science and scientists, between science and other modes of thought, between science as a description/explanation of nature and as an odious or merely dumbly callous philosophy, between science as a way of knowing and science as an intolerant truth, between the value of scientists as technicians and as scholars.

APES AND PEOPLE

Ultimately, there is no self-evident meaning in the structural similarity of chimp and human DNA, any more than there is in the structural similarity of our phlegm or our little toes. We know that we are similar to chimpanzees and yet distinguished from them.

Our place in nature is thus underdetermined by genetic data. To make sense of the data requires a biological eye and an anthropological mind, for its meaning—like the meaning of evolution a hundred years ago—is technologically constructed and ideologically situated.

Consequently, it should be no surprise that the collection of sophisticated genetic data today unfortunately all too frequently lends itself to remarkably unsophisticated interpretations. The divergence between the approaches of two of the founding fathers of molecular anthropology in the 1960s—Vince Sarich and Morris Goodman—illustrates this glaringly.

Both have demonstrated the genetic similarity of humans to chimpanzees and gorillas. Both have inferred the close relationship among these three creatures. But they differ on what that means.

Goodman feels that the overarching result should be a change in the formal classification of the apes, submerging all the living apes and humans within the family Hominidae, the great apes and humans within the subfamily Homininae, and chimps, gorillas, and humans within a tribe, Hominini. A more idiosyncratic interpretation of the molecular data further suggests to Goodman that chimpanzees and humans, but not gorillas, be recognized as constituting a subtribe, the Hominina.

"The traditional anthropological view emphasizes how very different humans are from all other forms of life," Goodman says. "In contrast, the view from molecular studies emphasizes how very much we hold in common . . . especially with chimpanzees, . . . with whom we share more than 98.3% identity in typical nuclear noncoding DNA sequences."

Of course it is a bit self-serving for someone studying molecules to privilege "the view from molecular studies," but that hardly gets to the heart of the matter. Is our frame of reference simply so myopic that all we get from the study of our genetic similarities to the apes is a proliferation of pseudo-Latin names?

Is that where the 98% similarity begins and ends? With the Hominini and Hominina?

It seems like a lot of effort for a fairly minuscule end product.

Vince Sarich takes a different view of the problem: the classification is trivial. For him, the issue is the evolutionary process that underlies the apparent pattern. Sarich quotes Thomas Huxley: "It would be not less wrong than absurd to deny the existence of this chasm [separating human and chimpanzee]; but it is at least equally wrong and absurd to exaggerate its magnitude, and, resting on the admitted fact of its existence, to refuse to inquire whether it is wide or narrow."

To which Sarich appends, "the chasm is indeed narrow, but very deep."

Sadly, genetics has only permitted us to glimpse the breadth of that chasm. They say that when the only tool you have is a hammer, everything tends to look like a nail—so it is not surprising that some geneticists would wish to ignore the chasm's depth, which they can't measure, and exaggerate the significance of its breadth, which they can.

But that tells us about those geneticists, not about apes and humans.

OH, NO—NOT ANOTHER PROJECT!

That's why I'm quite a bit skeptical of another "project" hoping to ride the coattails of the Human Genome Project's success to science nirvana. Given our 98–99% genetic similarity to the apes, why not sequence their genomes too and find out ultimately What Makes Us Human?

And thus, another group of molecular geneticists makes another self-interested appeal for lots of public money to undertake a "Human Genome Evolution Project."

The problem here isn't so much ethical as it is conceptual.

Recall, as noted earlier: *We don't map the genes for noses.* By this, I mean that in general the physical units of the body cannot be directly and specifically matched up with particular DNA segments. Just how to get a four-dimensional organism from the linear one-dimensional sequence of DNA, is entirely obscure.

Which is not to say that our big brains, penises, breasts, upright posture, and the rest of our biological specializations aren't ultimately programmed somewhere in our genes. Just, rather, that laying the human and chimpanzee DNA sequences side by side and examining their differences won't tell you what you want to know.

If what you want to know is why humans are human and not chimpanzees.

Sure, we will continue to find the phylogenetic relations of the apes to be quite ambiguous, with some DNA sequences, like mitochondrial DNA, appearing to link chimpanzees to humans; others, like the genes for involucrin and DRD-4, appearing to link chimpanzees to gorillas; and the DNA sequences of ancient retroviruses appearing to reveal a thoroughly intractable phylogenetic problem.

Sure, we will continue to find small differences between the genes of a chimpanzee and a human, and small differences in the structure of their chromosomes. We will even continue to find, as a group of researchers reported in 1998, small biochemical differences in cell structure. We will continue to find small differences between them wherever we look. That's what Edward Tyson found in 1699, and that's what we've been finding ever since.

But the suggestion that these particular DNA differences will be easily translated directly into something obvious or something crucial to the human condition flies in the face of what we know of genetics: that the connection between genotype and phenotype—between DNA and body—is very complex. Thus, as one advocate of the newest project gushes, "What happens if scientists identify a human gene that controls development of the larynx—a gene that might give chimpanzees the anatomy needed for speech?"

The geneticist is apparently anticipating Boo-Boo the chimp sitting up and saying, "Look, Ma, I'm talkin'!"

For one thing, the larynx alone won't do it; there's the tongue, the musculature, and the neural networks that control them. And for another, this view encapsulates a "macromutational" view of evolution—a view that is widely dismissed by evolutionary geneticists— to wit, that one or a few crucial changes to the DNA will have major effects on specific parts of the body, creating a "hopeful monster"— a talking chimp? a soccer-playing gorilla? a religious orangutan?— which will then become the progenitor of a new evolutionary lineage.

The problem, as we noted earlier in a slightly different context, lies with the assumption that highly pathological variation can be readily used to understand the range of normal difference and change. Genes with major effects on the body are invariably pathologies and also affect many different physiological systems. Populations of short-statured people like the central African pygmies are not simply groups of people that all have the gene for achondroplasia. Rather, they have "normal" genes for the low end of the range of human stature. If you went looking in their genomes for one gene with a major effect, with achondroplasia as your mind-set, you'd never find it. And we don't know what the "short stature genes" actually do, or where to find them, or how to identify them if we do find them.

So where does this leave us with the "Human Genome Evolution Project"? It's based on a problematic view of the evolutionary process, and would be designed to document once again precisely how narrow Huxley's chasm really is (the DNA sequence), and not how deep (the causal relationship between the genes and the bodies that make us and chimpanzees different). It's designed to give us more of what we already know, and not to deal with what would actually be interesting: the genetic physiology of humans and apes.

It is the genetic physiology of the apes that will ultimately allow us to learn why the AIDS virus appears to be so much less virulent in them than in us; a question that may have some impact on the treatment of the disease in people.

Twelve

SCIENCE, RELIGION, AND WORLDVIEW

SCIENCE IS A COGNITIVE system, a way to think about things. There are, of course, lots of ways to think about things.

Scientists are among the last great ethnocentrics in the modern world. And why not? They have a system that works; it produces technology. Why should they be humble? As Richard Dawkins, a leading spokesman for science, puts it, "The proof of the pudding is: When you actually fly to your international conference of cultural anthropologists, do you go on a magic carpet or do you go on a Boeing 747?"

The great paradox of modern science is that scientists are not trained to think about science; they are trained to *do* it, to carry it out. They are trained to use the machines with the flashing multi-colored lights, often in creative ways, to collect data—but not to think about where knowledge comes from, or the relationship between science and technology, or between scientific and nonscientific modes of thought, or even about the growth and development of their own field of science. These all fall within the domain of the humanities; generally a scientist is expected to pick them up osmotically, informally, passively. We all know about the ongoing explosions in science and technology, but we rarely hear about the quieter revolution that has been going on in the humanities about science. Sometimes scientists are vaguely threatened by the knowledge that someone is out there, studying them and thinking about them—as if they were the Yanomamo of the Amazon Basin or the !Kung San of Botswana.

But there are big questions now being asked of science that emanate from a paradox at the center of modern science. On the one hand, science is widely held to advance by a mechanism of "conjecture and refutation," in the words of philosopher Karl Popper, or in the more euphonious terms of immunologist Peter Medawar, "proposal and disposal." In other words, science advances by chucking out wrong ideas and keeping the relative few that work. On the other hand, science lays claim to being the domain of facts, of our positive knowledge of the natural world.

But how are we to reconcile science as disposal or refutation, which implies a great deal of nonfactualness within the province of science, to science-as-fact, which implies a claim to authority? Clearly, there is more here than simply the "discovery" of facts—there is a process by which a small class of ideas *become* facts. This process is partly based on ontologies—realities of nature—and also strongly rooted in a social matrix of research, publication, power, credibility, networking, strategizing, organizing, and propagandizing, which permit some facts to be recognized as such, and others to go unrecognized.

Thus one of the hottest intellectual areas today—some would say scorching—is the anthropology of science, the ethnographic study of how facts are made, through a negotiation between nature and the sociocultural web of science.

And there is an important corollary to this research. If facts are made rather than being simply discovered, then how do we know what the facts are at any point in time? The facts are, of course, what the men in white coats say they are.

But some of them are wrong. And we don't know which ones.

This makes the study of science very interesting, but it also is threatening, because it undermines science's claim to authority by virtue of facticity.

In their book *Fashionable Nonsense: Postmodern Intellectuals' Abuse of Science* (1998), physicists Alan Sokal and Jean Bricmont chide Bruno Latour, an influential French ethnographer of science, for failing to distinguish between a fact and the mere assertion of fact. The former is presumably something "out there"—the reality that science studies—and the latter is what scientists teach, what may or may not be true about the universe. But Sokal and Bricmont fail to address an important question: How do we tell them apart?

How can we distinguish in practice between what's "really" there and what the experts tell us is there—whether it's the number of chromosomes in a human cell, the motion of the solar system, or the nature of subatomic particles?

THE EUGENICS MOVEMENT

Or to take a concrete example—consider genetics in America in the 1920s. Pick up any textbook on the subject written in that era and you'll find a discussion of the merits of "eugenics," the idea that society could be improved through breeding better citizens. This not only implied involuntary sterilization of the worse citizens, but restricting the inflow of more bad citizens, which in effect meant the poor. The movement was popularized by nonscientists such as the influential Madison Grant, who (as noted in chapter 5) advocated the sterilization of "social failures" in his 1916 best-seller *The Passing of the Great Race*. And who were those "social discards" he deemed unworthy, unfit, and in need of gonadal surgery for the common good? Grant envisioned a process "beginning always with the criminal, the diseased, and the insane, and extending gradually to types which may be called weaklings rather than defectives, and perhaps ultimately to worthless race types."

Yup. "Worthless race types." Sure, you begin with the criminals, and you make your way quickly to pretty much anyone you don't like.

Grant's book was perceived as modern and scientific—he was friends with Charles Davenport, the leading human geneticist in America—and praised by politicians as diverse as his friend Theodore

Roosevelt (with whom Grant had helped found the New York Zoological Society) and Grant's admirer from afar, Adolf Hitler. (Hitler read the German translation of 1925. I'd like to think there were rather few other things he and Teddy Roosevelt agreed on.)

More than that, the book was reviewed and praised in the leading scientific journals. While popularized from outside the scientific community, eugenics was also very firmly rooted within it.

The eugenics movement spearheaded the drive to restrict immigration (enacted by Congress in 1924) and to sterilize the poor involuntarily (upheld by the Supreme Court in 1927, and likened to vaccination with the famous injunction that "three generations of imbeciles is enough").

The government believed it was acting according to the best modern scientific principles. And it was. The problem was, few people asked, "Who the devil are these scientists to be appropriating to themselves the decisions about whose stock is beneficial—who shall live, who shall die, who shall procreate?" To ask that question was to set yourself up as being anti-science, anti-modern, indeed, as anti-evolution—for eugenics donned the mantles of both Darwin and Mendel.

Consequently, Clarence Darrow was perceived as a traitor when he excoriated the scientific community for its eagerness to abrogate other people's rights. "Amongst the schemes for remolding society this is the most senseless and impudent that has ever been put forward by irresponsible fanatics to plague a long-suffering race," he wrote the year after defending John T. Scopes for teaching evolution.

Darrow had evolved from biology's champion to biology's basher in less than a year. Why? He was mortified by scientists pronouncing authoritatively on fields in which they had little training, little insight, and lots of arrogance. All they did was to bring the voice of authority to the aid of popular bigotry.

And they refused to acknowledge it, because they were speaking the facts.

The eugenics movement died with the onset of the Depression. After the Crash, geneticists belatedly came to appreciate that social worth and genetic endowment might not be so closely connected.

But there was no great scientific discovery involved—simply a grow-ing appreciation for human rights, an appreciation that they saw erod-ing quickly in Germany. And American geneticists were actually quite conflicted about it.

One prominent geneticist believed that the Germans were "beating us at our own game." Others were less outspoken but still conflicted.

In 1934, colleagues and students of the leading geneticist in Ger-many, Eugen Fischer, honored him with a special issue of the *Zeit-schrift für Morphologie und Anthropologie*. In their preface, the editors glowed about the new government firmly in place: "We stand upon the threshold of a new era. For the first time in the history of the world, the führer Adolf Hitler is putting into practice the insights about the biological foundations of the development of peoples—race, heredity, selection. It is no coincidence that Germany is the locus of this event: German science provides the tools for the poli-tician."

In Fischer's honor, the two leading human geneticists in America contributed papers to that volume—Raymond Pearl of Johns Hop-kins, and Charles Davenport. Neither lived to see the end of World War II. Wouldn't it be interesting to know how they felt in retrospect, about having their own words appear behind those just quoted?

To complete the story—the preface was written by Otmar Freiherr von Verschuer, a student of the honoree Eugen Fischer, and quite obscure except for the fact of having been the academic advisor of Josef Mengele, the infamous Auschwitz camp doctor. And Fischer himself applied for and received full Nazi party membership in 1940.

THE LESSON

What are we to make of the eugenics movement? One modern ge-neticist writes for his student readers: "As is often the case in science, geneticists have become much more humble about their understand-ing of their subject as they realise how little they really know. Eugenics was based on ignorance and prejudice rather than on fact; a science with these at its centre was bound to die."

There's a lot to disagree with here.

For one thing, it's one of rather few places you may ever see the word "geneticists" modified by the word "humble." (After all, in July 2000, when geneticist Francis Collins announced at the White House that the Human Genome Project had completed 90% of its ambition to sequence the human genome, he proclaimed, "We have caught the first glimpses of our instruction book, previously known only to God." So it was not surprising to hear President Clinton echo the sentiment, "Today we are learning the language in which God created life." Those humble geneticists!)

And for another thing, it still doesn't indicate how to tell "ignorance and prejudice" from "fact," nor does it account for why the leading scientists of the day were strikingly and pathetically unable to do so themselves. The point is that geneticists did not—until very late in the game—say, "Hey, this is based on ignorance and prejudice, not on fact! It's bound to die!"

They couldn't tell the fact from the assertion of the fact.

And whether it was "bound to die" or not is beside the point; it *did* die, but not before harming many people (while it would be an overstatement to blame Nazi policies on geneticists, books by American scientists with titles like *Sterilization for Human Betterment* widely extolled the virtues of social surgery and bragged about the involuntary sterilization of nearly 10,000 poor people in California alone).

The relationship between reality (the facts) and the views of the experts (the assertion of fact) is dependent fully upon the wisdom and goodwill of those experts. They control the means for telling the facts from the nonfacts; if you are skeptical about what you hear about subatomic particles, after all, you can't exactly go out and build your own cyclotron. And it is, of course, in the interests of those experts to assure the rest of us that there is nothing to worry about. Thus, the "real world" of science—the conflicts of interest, the power struggles, the politics, the stupidity, the arrogance, the lapses of integrity—are not to be studied or dwelt upon. They are simply defined out of existence or rendered invisible—science is the study of the facts of nature.

But we need to do better than the geneticists of the 1920s. We need to understand both nature and the relationship between nature and representations of it. We need to know how things become facts, so that we can tell the facts from the crap without having to wait fifty years to see what future scientists think about it.

But the ways in which facts are negotiated between nature and culture, the way in which those facts achieve facticity, are not part of the formal training of scientists; they are simply taught to believe they are reading nature.

How sad!

It almost makes you wonder how science manages to progress, in spite of the training of scientists.

RELATIVISM AND SCIENCE

So let's return to Richard Dawkins's 747 and flying carpet. Does anyone really think that you can travel to conferences on a flying carpet? The 747 works; it's real; it gets you where you're going.

But let's turn the question around. Does the fact that a 747 works mean that the scientific view of the world is superior to others?

A 747 is not science, even in microcosm. It is technology, a *product* of science. And as anthropologists since Franz Boas have noted, technology invariably constitutes a series of trade-offs. With the wonders of the 747 come the wonders of lost luggage, cramped seats, no escape from the obnoxious strangers around you, and the fear of bombs on board. Sure, the Yanomamo don't write books on personal computers with Pentium©-based processors. But by the same token, they don't worry about carpal-tunnel syndrome, system crashes, the Michelangelo virus, or any of a host of other anxieties that accompany the technology. The disadvantages to our world of televisions and nuclear weapons need hardly be tallied, although their utility is equally obvious.

Ultimately, technology is part of a cultural system, and as it changes, so do other elements. And it is never clear whether the result is improvement or decadence. Generations of social philosophers have argued both ways, for example, Francis Bacon glamorizing the technological society in the seventeenth century and Jean-Jacques Rousseau deriding it in the eighteenth century. Ultimately, though, whether technology is good or bad, we are stuck with it.

Which brings us back to the contrast between a 747 and a flying carpet. The former is a piece of technology, the latter a piece of mythology. We can compare technologies and argue for the superiority of one over the other (although it is unclear whether lives overall are improved or degraded by it). We can compare mythologies, but there is no scale on which to tell whether one is superior to another. But to compare technology to mythology is ridiculous, even perverse.

The biologist apparently believed he was refuting cultural relativism, a principle that arose in the early twentieth century in the wake of the exploitation, destruction, and genocide that accompanied colonialism. Often it was justified on the grounds that Euro-American society had progressed further along a cultural or technological continuum and was therefore better than the cultures and peoples Euro-Americans were wiping out.

It was a convenient rationalization, but it turned out to be difficult to prove, as ethnographers documented the depth of local knowledge of the environment, the complexities of diverse languages, and the richness of non-Western lifeways. Cultural relativism became an intellectual principle governing the comparison of cultural systems—that they were different from one another, but could only be ranked in very narrow fashions.

Cultural relativism is occasionally confused with *Star Trek*'s "prime directive" of nonintervention, but isolationism was never part of the doctrine of cultural relativism. Indeed, its most vocal exponents were anti-Nazi activists. Cultural relativism is also occasionally confused with a nihilistic philosophy of moral relativism in which "anything goes." But this is the very opposite of cultural relativism, which recognizes that we all live within a cultural system that distinguishes right from wrong—and although those distinctions may vary across societies, one is obliged to live within the local set of norms or risk the imposition of sanctions.

Science is a part of our cultural system. It is a way of seeing the world and a way of finding out more about the world. To paraphrase

Peter Medawar again, science is a means by which we analyze the many things that *might* be true about the universe and pare them down to the relatively few that probably *are* true. In this way, we gain positive knowledge, which we then apply to generate new technologies and mastery over nature.

The concept of "mastery over nature" is itself culturally loaded. It is certainly tied into assumptions of social hierarchy, and as Mary Midgley has shown, to images of sexual violence—science as male, nature as female.

But why do we even believe we need to dominate nature? Why do we need to know exactly how things work? We take these assumptions for granted because we have inherited a cultural view that takes positive knowledge of nature as an ultimate goal, rationalism as a guide to life, and technology as a mode of existence. That is certainly a valid view. But is it the only acceptable one? Is it necessarily the best one?

Let's examine rationalism. Scientists perennially fret about the irrationality of the common person (a recent example is the late Carl Sagan's *The Demon-Haunted World*). They deplore the fact that almost all newspapers carry horoscopes, the fact that books like *Chariots of the Gods* and *The Celestine Prophecy* can be huge best-sellers, the fact that Americans are so religious and so resistant to accepting the theory of evolution. To a class of scientists, these nonscientific modes of thought are threatening and to some extent foolish.

Contrasting what he calls "transcendentalism" with science, entomologist E. O. Wilson, the father of sociobiology, notes that the former is "full and rich" and also "easier to follow"—thus it "continues to win the heart" even as science is "winning the mind." And with more than a bit of condescension, he observes:

> Millions . . . feel otherwise adrift in a life without ultimate meaning.
> . . . They enter established religions, succumb to cults, dabble
> in New Age nostrums. They push *The Celestine Prophecy* and
> other junk attempts at enlightenment onto the best-seller lists.
>   Perhaps, as I believe, it can all eventually be explained as brain
> circuitry and deep, genetic history. But this is not a subject that
> even the most hardened empiricist should presume to trivialize.

This contrast was made more sensitively and cleverly by Cole Porter in a song he wrote for *Silk Stockings,* the musical version of the classic Garbo movie *Ninotchka.* The lovely Soviet empiricist/sociobiologist sings of love:

> When the electromagnetic of the female
> meets the electromagnetic of the he-male,
> if right away
> she should say
> "This is *the* male"—
> It's a chemical reaction, that's all.

But her dashing counterpart doesn't buy it. Challenged to produce his own theory, he responds, "My theory is that there is no theory" and croons:

> I love the looks of you,
> The lure of you.
> I'd love to take a tour of you.
> The eyes, the arms, the mouth of you;
> The east, west, north, and the south of you.
> I'd love to gain complete control of you
> And handle even the heart and soul of you.
> So love at least a small percent of me, do,
> For I love all of you.

Guess who wins.

Of course, love and mystico-religious experiences are electrochemical. But water is merely hydrogen and oxygen. Yet that knowledge helps us neither to slake our thirst after a long mountain hike nor to cope with another year of drought and the threat of the crops failing; nor does it permit us to experience the mysteries of baptism or to rebuild our lives after seeing our homes wrecked in a flood.

Just hydrogen and oxygen. But that knowledge is trivial; the extraordinary thing is why the reductive scientific attitude overvalues itself and laments everyone else's shortsightedness.

Wilson believes that "we evolved genetically to accept one truth and discovered another," thus issuing a general absolution to science for its overarching failure here. But his thought about the apparent conflict between science and religion fails to come to grips with what anthropology has learned in the past century. It may be worth noting that Wilson also blames the failure of his joyless reductionism to penetrate the masses deeply on "ignorance of the natural science by design[, . . .] a strategy fashioned by the founders [of modern social science], most notably Emile Durkheim, Karl Marx, Franz Boas, and Sigmund Freud." But that's a bum rap, for none of them was anti-science: Freud had a Viennese medical education, the envy of any in contemporary America; Boas had a doctorate in physics; Marx read and appreciated far more of Darwin's work than Darwin did of his; and Durkheim based his famous model of society (organic solidarity) on a biological analogy.

They understood what the scientists of their day were saying about the important things in human life, and they simply opposed it.

Anthropologists of the nineteenth century tended to see a progression of human thought from magic or superstition to religion to science, with the rationality of the scientific modern age ultimately coming to supersede and replace the stupidities of earlier times and cultures. But we abandoned that way of thinking in the twentieth century. Why? In the first place, it is not at all clear that magical or superstitious thought has been superseded by religion, much less religion by science. People today are no less superstitious than they were hundreds of years ago; that kind of belief simply emerges in particular contexts—sports and gambling, for example. By the same token, Americans in the paramount technological society of the modern age are extremely religious.

Is it, as the writings of some scientists imply, because those people are stupid?

There is a different explanation. Magic and religion are with us, just as strong as ever. Even among scientists. It is the exaggerated claims of scientists, and their often arrogant misunderstandings of human behavior and society, and of the domain of science itself, that are the problem.

What kind of scientific rationalities should we make people follow?

Conceptions of what scientific rationalities should govern modern lives are invariably strongly culture-bound. The American geneticist Charles Davenport in 1911 envisioned a eugenically scientific society in which people would "fall in love intelligently." Not only were he and his followers attempting to impose a cold rationalism upon a fundamentally inappropriate substrate, but their idea of scientifically intelligent mating essentially excluded any partner who was not wealthy, abstemious, and Nordic. Clarence Darrow condemned Davenport's elitist scientism as the product of an "age of meddling, presumption, and gross denial of all the individual feelings and emotion."

The idea of people leading all aspects of their lives by scientific principles has been attractive to utopian writers, but most attempts over the past two centuries to implement reforms in that direction have failed extensively. For example, our numerically bizarre time-keeping system (the 24-hour day and 60-minute hour) is an inheritance from ancient Babylonian arithmetic systems, and our ridiculous 7-day week and 12-month year from their astrological cycles. Wouldn't a rational, decimal time-keeping system be superior? One-hundred-minute hours, ten-hour days, ten-day weeks, and ten-month years. Yet attempts to impose such reforms were unsuccessful following the French Revolution's attempt to establish a scientific society; or following the Soviet Revolution's attempt to do the same. Britain adopted a decimal currency only within the past generation, and its scientists aside, the United States has steadfastly resisted the metric system of weights and measures.

Many rational scientific systems for running people's lives have, in fact, been proposed, but what is considered in a given time and place to be a rational scientific thing to compel people to do may not necessarily even be good science, let alone display any respect for human rights. To cite only the most flagrant example, the Third Reich's policy of *Rassenhygiene* offered what purported to be a rational, scientific solution to the problem of large numbers of inferior and undesirable people.

This does not mean that the imposition of scientific rationality in general life is wrong; only that one needs to consider its track record before placing one's bets. Utopian scientific visions notwithstanding, science fiction plots (e.g., *Star Trek*) ridicule hyperrationalism as less than human, if not fundamentally contrary to human nature. Fun, of course, is nonscientific and irrational, but indelibly human.

It is not simply, as E. O. Wilson argues in *Consilience,* that we evolved to be irrational, and then science discovered a rational approach to the universe that is destined to overlay it. Rather, both together constitute human nature, and neither is more or less human than the other. But one without the other constitutes an incomplete, inhuman condition. Perhaps, indeed, the sciences can learn about human nature from the humanities.

What the humanities tell us is that science in modern life is not merely rational thought but bureaucratically organized rational thought. In other words, science is both about the universe and about scientists; it is both a rational and a social process. Consequently, scientific statements encode two messages that are not strictly compatible: accuracy and authority. Scientific statements strive to be accurate, and often succeed, but are *always* authoritative. Why? Because they are made by scientists. This creates a problem. If scientific statements are equally authoritative whether or not they are accurate, how do we know when to believe them?

The authority of scientists has been exploited for generations by Madison Avenue. Whether it was that smoking Brand X is actually good for you or that eating one or another high-fat, low-fiber, sugary food is healthy, the ability to get "9 out of 10 doctors" to say it has always been recognized as a selling point. The authority that comes with statements by scientists comes with responsibilities—the responsibility to get it right, to quickly censure those who get it wrong in the name of science, and to acknowledge the consequences of misleading the public.

Unfortunately, it is no easier to get the average scientist to accept responsibilities than it is to get the average four-year-old to accept responsibilities. Scientists wish to speak with authority, to be able to say outrageous things, and to let outrageous statements by their colleagues stand unchallenged or minimally challenged and still have their next utterance carry the authority of science too. That is not science—that is scientism, an uncritical faith in science and scientists. And that is what is being challenged here.

In addition to authoritative statements, science of course also makes accurate statements. Science, after all, strives to comprehend the universe. As any scientist will tell you, it consists of formulating hypotheses, testing them against empirical data, falsifying those that are inaccurate, and linking them together as explanations or theories. In this way, of course, science works—we understand things better now than we did a hundred years ago, and expect to understand them even better a hundred years from now.

But if, as generally formulated in the latter half of the twentieth century, the principal mechanism of scientific advancement is the falsification of hypotheses, it is difficult to equate "scientific" statements with "accurate" statements. After all, most hypotheses concerning a particular phenomenon will ultimately be falsified; this will lead to greater accuracy in the long run, but it also means that most ideas about it were wrong, although no less scientific for that reason.

Scientists are routinely taught, for example, about Mendel's discoveries of the basic principles of heredity. Sometimes they are taught that he reported his data selectively, or even fudged them a bit, but of course one should hesitate to judge 1860s research by today's standards. What scientists *aren't* taught is that Mendel's scientific contemporaries certainly had their own ideas about heredity. They weren't dunces; their ideas were scientific, interesting, and nevertheless wrong.

Thus, science is generally not accurate, except in the long run. Scientific statements are routinely falsified. In other words, most scientific statements turn out to be inaccurate, and rather few actually turn out to be accurate. That is simply a consequence of the way science operates. The 747 flies because of thousands of other aircraft ideas that didn't.

So why should the 747 be entitled to bragging rights over flying carpets, which have been the subject of far fewer failed ideas, designs, and experiments?

There is one more important difference between the ideas on which scientifically engineered aircraft are based and the idea of flying carpets—namely, their purpose. Science is constituted as a series of explanatory stories about the universe. Explanatory stories are, like any other stories, related through the medium of language, using metaphors to communicate meanings—the selfish gene, the Big Bang, the genetic code. These cannot be literally true: Genes are inanimate and can no more be selfish than they can be loyal, ferocious, or pious; there could not have been any bang at all without an atmosphere to transmit sound, and in the absence of matter there could not have been atmosphere; and the genetic "code" was a metaphor devised in 1944 by the physicist Erwin Schrödinger, before molecular genetics was even invented.

So in the anthropological usage, stripped of the pejorative connotations of the common use of the term, science is constituted by myths—linguistic stories purporting to explain things. Every culture has its own. But those of science are a bit different. Scientific stories have but a single goal, to relate as accurately as possible (given the constraints of knowledge and language) the phenomenon in question.

If accuracy is the scale by which to compare myths, certainly science's myths win. But that is an obviously rigged game, using science's criterion to evaluate its alternatives. The simple fact that science is ignored or rejected by a significant number of otherwise modern people should suggest that they have other criteria that matter to them, even if science doesn't.

Consider, for example, some of the many things that the scientific myth of human evolution doesn't address. Affirmation of a sense of individual worth or importance; living right, by codes of morality or standards of conduct; the evocation of strong emotional bonds of solidarity to a community; or simply how to get through the day and feel good.

This is, of course, not to defend creationism, by which fundamentalist Christians seek to subvert science education in America. It is to suggest, however, that the rejection of science by ordinary citizens is a reflection of the failings of scientists, not of the failings of those ordinary citizens. Representing creationism as science is quite simply a fraudulent claim. But it is crucial to note that science provides a restricted set of answers to a very large set of questions that people in all cultures have; and that science may easily be judged inadequate if we look beyond the sole criterion of empirical validity. Evolution provides the most empirically valid explanation that we have for the present existence of life. Period.

But why should it really matter whether we are descended from arboreal hairy primates or not? Either way, we are still human, and engaged in the construction of a civil and just society. Either way, you still have to earn a living, put food on the table, cope with injustice and suffering, and find meaning in your life.

The reason it matters to so many people is that scientists have made it matter, and they've done so in the worst possible way. They've taken a proposition that does not matter much and has little effect on the lives, cares, and worries of everyday people—"We are descended from apes"—and stretched that into a series of additional propositions, often both authoritative and odious.

Thirty years ago, in a widely read scientific-philosophical work called *Chance and Necessity,* the French molecular biologist Jacques Monod argued that evolution shows life to be meaningless. While this might easily be dismissed as "Sartre among the test-tubes," it carried the authority of science, because a prominent scientist wrote it. Is the proposition true? Perhaps, but there is no way to know. There is no class of data to be collected that would indicate whether life is meaningful or not. It is not a scientific proposition.

More than that, it is a distasteful proposition. Bluntly put, people care more about whether their life has meaning than they do about whether they came from apes. If you tell them that science shows life has no meaning and that we came from apes, it is not surprising to imagine that they would reject both scientific propositions. In fact, it is pretty dumb to think otherwise.

The problem is that science is very good at answering questions people don't care about. To the extent that physics aids the technology that allows you to reheat frozen food in a few minutes, it is obviously useful. But using technology derived from it, and caring about it, are different things. The things that people care about tend to be the things outside the domain of science—What is death? Will I always be able to take care of my children? Why do good things happen to bad people? How can I be happy?

All humans care about these things. All cognitive systems provide answers for them. In addition, they provide explanations for how humans and the world they live in came to be—as the scientific myth does. And more than that, other myths explain not simply how we came into existence, but *why*.

And science doesn't.

Science explains how we came to exist more accurately than does any other myth. By its own criterion, it is therefore the best explanation. But it is an answer to a relatively small and trivial question. Science tells us that we are descended from apes, a fact that affects people's lives and minds minimally, if at all. On the other hand, science says nothing about whether the cosmos is ultimately benevolent or just. The perpetual crisis in science education is largely the result of a consistent failure of scientists themselves to be educated about what they do and its implications.

Richard Dawkins writes in *River Out of Eden:*

> [I]f the universe were just electrons and selfish genes, meaningless tragedies . . . are exactly what we should expect, along with equally meaningless *good* fortune. Such a universe would be neither evil nor good in its intention. It would manifest no intentions of any kind. In a universe of blind physical forces and genetic replication, some people are going to get hurt, other people are going to get lucky, and you won't find any rhyme or reason in it, nor any justice. The universe we observe has precisely the properties we should expect if there is, at bottom, no design, no purpose, no evil and no good, nothing but blind, pitiless indifference.

And for good measure he adds, "DNA neither knows nor cares. DNA just is. And we dance to its music."

Well, maybe we do and maybe we don't. Since the author is not an expert on molecular genetics, we may consequently take his musings about DNA with a grain of salt, falling within the domain of folk heredity. But what about his dour view of the universe? Here Dawkins presents a consistency argument, not a test of an hypothesis.

After all, the universe also has precisely the properties we would expect to find if it *were* benevolent and designed, *and we simply didn't understand it,* lacking the key to its pattern. If the only language you speak is Greek, other languages may sound like random noise, like "bar, bar, bar," which is why the Greeks called non-Greek speakers "barbarians." But that's a statement about the limitations of the Greeks, not about the other peoples.

Random noise may be random noise or it may simply be stuff you don't understand, which consequently looks or sounds like random noise. The history of modern science, after all, is about the discovery and imposition of order on what formerly looked like chaos. Perhaps ultimately there is just chaos, but Dawkins's assertion about it is no more than that, an assertion.

Dawkins's interpretation of the universe *might* be true, but again, since there is no positive knowledge we can acquire, no controlled set of data we can collect that would indicate whether it is in fact *likely* to be true, we are obliged to identify the statement as nonscience.

The scientist is, of course, welcome to his opinion. It is not, however, the case that his opinion about this is more scientific than any other. Indeed, since it reflects an inability to tell science from nonscience, it might actually be regarded as *less* scientific than any other. The important criticism, however, lies in the implications of teaching such philosophy as if it constituted science, indeed as if it constituted the theory of evolution.

The scientist says: "Science has explained many things about the universe. Your life has no meaning. Have a nice day." And then he is surprised and appalled at the public rejection of that philosophy. If the goal of science is to make people miserable, then Dawkins and his gloomy philosophy would seem to be the ideal evangelical tool.

But for those of us who think that perhaps people do have the right to be happy (or at least, as Thomas Jefferson believed, the right to pursue happiness), it is an impoverished and unfulfilling worldview. Small wonder it is so unpopular! Small wonder that people would rather derive pleasure from the comforting inanities of *The Celestine Prophecy.*

ETHNOCENTRISM AND THE FUTURE OF SCIENCE

Science attacks complex, integrated mythologies with a unidimensional substitute: one that is simply more accurate about mundane, mechanistic things. Science thus sets for itself a single goal—empirical accuracy—and relentlessly attempts to supplant other systems of belief on that basis, leaving an intellectual vacuum in its wake in many other ways. In other words, science does chicken right but frequently leaves diners bored and nutritionally imbalanced.

There is considerable ideological baggage associated with compelling nonscientists to think as scientists do—or at least as the most cynical scientists do. It may be worth reflecting upon the implications of denying people the other qualities that nonscientific origin myths provide. Science has only recently and marginally come to consider its responsibilities, and in our culture, the responsibility attached to telling people authoritatively that they are unimportant in the universe is great.

And it may be worth considering as a question of scientific ethics whether, in an insecure world, science may actually be doing more harm than good to people in actively undermining their images of self-worth in a benevolent cosmos. Perhaps, rather than asserting the authority of science aggressively in opposition to whatever else is popularly and ignorantly believed, a humanistic, anthropological approach to science, as a set of ideas about the universe arising from a particular historical development and serving a particular function, may be a more effective way of getting its central messages across. Teaching *how scientists think about science* is fundamentally different from compelling people to think the way scientists do (or the way they are supposed to think).

Is it fair, then, for scientists to tell citizens of our nation, or of *any* nation, "You are wrong and we are right"? This is an ideological position widely held by powerful societies when confronted with the beliefs of less powerful societies. It is the rawest form of ethnocentrism.

Is ethnocentrism wrong? It is certainly an ineffective way of getting people to like you or to respect you. It is an effective way of getting people to fear and resent you. If science adopts such a stance, it is not terribly difficult to predict the probable reaction.

A reasonable alternative to the ethnocentric presentation of science is to present science in an anthropological framework. In other words, to present science not so much as the one true answer in opposition to the many false ones (which has a familiarly evangelical ring to it) but as *an* answer constructed within a particular cultural framework, satisfying certain criteria well (notably empirical validity), but other meaningful criteria poorly.

Humanistic ideas intersect with scientific ones at almost every turn in the study of humans. We cannot escape them, and it is false to suggest that they can successfully be ignored in the modern era, when social and political problems constitute the bulk of unsolved problems in contemporary life. We now acknowledge that the study of human biology carries responsibilities that come with the authority of speaking as a scientist. Science has justified racism, but it should not. Science has justified colonialism, but it should not. Science can make people's lives miserable, but it should not.

Decades ago, José Ortega y Gasset bluntly criticized the radical specialization of modern scientists as comprising a new class of "learned ignoramus . . . ignorant, not in the fashion of the ignorant man, but with all the petulance of one who is learned in his own special line." Anti-intellectualism, after all, comes in many forms and imperils not simply science but the entire scholarly community.

Consider, for example, the misrepresentation of creationism as science (from outside of science), and the misrepresentation of genetics as a social panacea (from within science). Creationism is insidious, because by misrepresenting itself as science it promotes ignorance. But neohereditarianism, also misrepresenting itself as good science, is insidious as well. Is either threat greater than the other? As far as I am aware, no one has ever been killed or involuntarily sterilized in the name of creationism. From the standpoint that ignorance is bliss, and death probably is not, we might consequently do well to be at least as skeptical of the judgments of scientists as of anti-scientists, at least where people's welfares and lives may be concerned. Indeed, until such a time as modern social ethics becomes a required subject for science students, we might do well to be even *more* skeptical of the social judgments of scientists than of other citizens. As one commentator has noted, "[W]ithout any reference to the interaction between science and society, [science education] will tend to produce naïvete, xenophobia, and intellectual arrogance."

Because of the strides made by science, it is now possible for poor, unethical, or pseudoscientific judgments to affect more people's lives more catastrophically than ever before. Science thus needs the humanities now more than ever.

If the twentieth century was the century of modern science, the twenty-first will be the century of modern scientific responsibility. Science will have to be uncoupled from the colonialism and ethnocentrism of the modern era, which in turn implies a somewhat different approach to the training of scientists. The efficiency with which we can actually use science to help solve problems of health and welfare in a global community may depend upon how effectively we are able to accomplish this.

As a mediator of different and incompatible bodies of knowledge, anthropology has an important role to play in the relation of science to modern society. To the extent that genetics represents the molecular science of heredity, the "molecular anthropology" at the heart of this book should be a central case for understanding the ambiguous nature of science's relationship to society. Science is seen as both authority (epitomized in the benevolent Dr. Einstein) and sociopathy (epitomized in the deranged Dr. Frankenstein). The reason for this lies not simply in the failures of the masses (to which science ordinarily likes to attribute it) but also in the failures of scientists, whose training generally does not include a curriculum on the humanistic and social aspects of what they do.

Thus, Patrick Tierney's recent book *Darkness in El Dorado: How Scientists and Journalists Devastated the Amazon* struck a nerve with anthropologists by exposing the scientific, economic, and sexual exploitation of the Yanomamo of the Amazon basin by ambitious and ruthless scientists. Unfortunately, *Darkness in El Dorado* fails to live up to the standards demanded of an exposé. It accuses geneticists of starting a measles epidemic with their vaccination program, but it would seem that the program was, in fact, initiated at least in part as a humanitarian gesture and that measles preceded the scientists. And with forty years of hindsight, it is a bit unfair to second-guess each decision made in the field at the time.

But the book does raise some important issues about the responsibilities of scholars to the communities they study, ethical issues that anthropologists have been grappling with for decades. In the context of an archaic anthropology, when much of the field strove to be purely objective and scientific (in the sense of what is sometimes called "potted philosophy of science"), giving something back to tribal peoples would have been as scientifically senseless as chemists giving something back to boron. But that isn't the case today.

Anthropologists have been debating ethics for as long as any scientists and speaking out on behalf of the powerless. The very first academic anthropologist, Edward Tylor of Oxford, called the field "essentially a reformer's science" over a hundred years ago. And there is hardly a more beloved target of political conservatives in academia than Margaret Mead. Anthropology is very self-consciously a science of the heart as well as of the head.

So, ultimately, one has to be slightly uncomfortable with the fact that Yanomamo DNA sits in a refrigerator at Penn State, exclusively for the benefit of scientists in perpetuo, as Tierney notes, and is not under the control of the Yanomamo themselves. That is a modern problem, one that requires some thought in the context of balancing human rights and welfare against the advancement of science (and the advancement of scientists).

"Scientism is the secular religion of the day, and what I have long feared has come to pass: increase of knowledge unaccompanied by increase of understanding," Ashley Montagu wrote presciently nearly half a century ago to his friend Theodosius Dobzhansky, who understood the value of anthropology for genetics.

In sum, the place of the human species in the natural order is predicated on the place of the chimpanzee, and is consequently a contested site on the boundary of animalness and godliness, beast and angel. When genetics provides information bearing on that question, it does so within the contexts of both the scientific study of heredity and the study of human systems of meaning. It cannot be divorced either from its intellectual contexts, social and philosophical implications, or the responsibilities of scientists.

And more than that, it provides an entry for anthropology into molecular studies to create a truly interdisciplinary research area.

Molecular anthropology shows in microscosm how we can integrate science into our contemporary culture and acknowledge science as a significant part of it, while at the same time valuing other cultural subsystems and other kinds of knowledge. The other kinds of knowledge are as important for the proper interpretation of scientific data as science is for "modern life." Humanistic knowledge is thus at least as crucial to the scientist as scientific knowledge is for the masses. Ultimately, this hybrid field permits us to bridge "the two cultures" and give a broader and more effective base to understanding of the molecular basis of human existence.

NOTES AND SOURCES

INTRODUCTION

C. P. Snow, *The Two Cultures and the Scientific Revolution* (New York: Cambridge University Press, 1959). First published in preliminary form in the *New Statesman,* October 6, 1956.

Richard J. Herrnstein and Charles Murray, *The Bell Curve: Intelligence and Class Structure in American Life* (New York: Free Press, 1994).

CHAPTER I

For an overview of primate evolution and systematics, see John G. Fleagle, *Primate Adaptation and Evolution* (San Diego: Academic Press, 1988; 2d ed., 1999).

On the work of Sarich and Wilson, see V. M. Sarich, "Appendix: Retrospective on Hominoid Macromolecular Systematics," in *New Interpretations of Ape and Human Ancestry,* ed. R. L. Ciochon and R. S. Corruccini (New York: Plenum Press, 1983), 137–50.

On weird phylogenies based on molecular data, see G. J. P. Naylor and W. M. Brown, "Structural Biology and Phylogenetic Estimation," *Nature* 388 (1997): 527–28; J. P. Curole, and T. D. Kocher, "Mitogenomics: Digging Deeper with Complete Mitochondrial Genomes," *Trends in Ecology and Evolution* 14 (1999): 394–98; W. W. de Jong, "Molecules Remodel the Mammalian Tree," *Trends in Ecology and Evolution* 13 (1998): 270–75; D. Graur, L. Duret, and M. Gouy, "Phylogenetic Position of the Order Lagomorpha (Rabbits, Hares, and Allies)," *Nature* 379 (1996): 333–35; D. Graur, W. A. Hide, and W.-H. Li, "Is the Guinea-pig a Rodent?" *Nature* 351 (1991): 649–51.

For Sarich's view from retirement, see V. M. Sarich, "Molecular Clocks: Then and Now" (paper presented at the Dual Congress of the International Association for the Study of Human Paleontology and the International Association of Human Biologists, Sun City, South Africa, 1998).

On early studies of the apes, see R. M. Yerkes and A. W. Yerkes, *The Great Apes: A Study of Anthropoid Life* (New Haven: Yale University Press, 1929).

For Andrew Battell, see Samuel Purchas, *Hakluytus Postumus, or, Purchas His Pilgrimes, Contayning a History of the World in Sea Voyages and Lande Travells by Englishmen and Others* (1625; Glasgow: James MacLehose, 1905), 6: 397–99.

Nicolaas Tulp, *Observationum medicarum libri tres . . .* (Amsterdam: Apud Ludovicum Elzevirium, 1641).

Edward Tyson, *Orang-outang, sive homo sylvestris, or, The anatomy of a pygmie compared with that of a monkey, an ape, and a man to which is added, A philological essay concerning the pygmies, the cynocephali, the satyrs, and sphinges of the ancients: wherein it will appear that they are all either apes or monkeys, and not men, as formerly pretended* (London: Thomas Bennett & Daniel Brown, 1699). Facsimile edition with an introduction by Ashley Montagu (London: Dawsons of Pall Mall, 1966).

Willem Bosman, *A New and Accurate Description of the Coast of Guinea, Divided into the Gold, the Slave, and the Ivory Coasts* (London: J. Knapton, 1705).

Speaker chimpanzee: "Animalis rarioris, chimpanzee dicti," *Nova Acta Eruditorum* 8 (1739): 564–65, an anonymous report attributed to Hubert Gravelot by G. Barsanti, "Les singes de Lamarck," in *Ape, Man, Apeman: Changing Views since 1600: Evaluative Proceedings of the Symposium . . . Leiden, the Netherlands, 28 June–1 July 1993*, ed. Raymond Corbey and Bert Theunissen (Leiden: Dept. of Prehistory, Leiden University, 1995).

Linnaeus's *System of Nature:* [Carl von Linné], *Caroli Linnæi Systema naturae, sive, Regna tria naturae systematice proposita per classes, ordines, genera, & species* (Leiden: Apud Theodorum Haak, Ex typographia Joannis Wilhelmi de Groot, 1735); 10th rev. ed., *Caroli Linnæi Systema naturæ per regna tria naturæ secundum classes, ordines, genera, species, cum charateribus, differentiis, synonymis, locis* (2 vols., Stockholm: Impensis L. Salvii, 1758–59). And see also Lisbet Koerner *Linnaeus: Nature and Nation* (Cambridge, Mass.: Harvard University Press, 1999).

Thomas Pennant, *Synopsis of Quadrupeds* (Chester, Eng.: J. Monk, 1771).

Johann Friedrich Blumenbach: *On the Natural Variety of Mankind,* 3d ed. (1795), in *The Anthropological Treatises of Johann Friedrich Blumenbach,* trans. and ed. Thomas Bendyshe (London: Longman, Green, 1865).

George Gaylord Simpson, *The Principles of Classification and a Classification of Mammals. Bulletin of the American Museum of Natural History* 85 (New York: American Museum of Natural History, 1945).

Julian Huxley, "Evolution, Cultural and Biological," *Yearbook of Anthropology* 0 (1955): 2–25. (This is formally listed as volume zero because the *Yearbook of Anthropology* evolved into *Current Anthropology* shortly thereafter.)

On genes for ribosomal RNA, see I. L.Gonzalez, J. E. Sylvester, T. F. Smith, D. Stambolian, and R. D. Schmickel, "Ribosomal RNA Gene Sequences and Hominoid Phylogeny," *Molecular Biology and Evolution* 7 (1990): 203–19.

"In a single recent book on the great apes . . .": *The Great Ape Project: Equality beyond Humanity,* ed. Paola Cavalieri and Peter Singer (1993; 1st U.S. ed., New York: St. Martin's Press, 1994).

On DNA cologne, see Jonathan Marks, "Arrivaderci, Aroma: An Analysis of the New DNA Cologne," in *The Best of the Annals of Improbable Research,* ed. Marc Abraham (New York: W. H. Freeman 1998), 121–22.

On the number of human chromosomes, see M. Kottler, "From 48 to 46: Cytological Technique, Preconception, and the Counting of Human Chromosomes," *Bulletin of the History of Medicine* 48 (1974): 465–502.

CHAPTER 2

On differences between human and chimpanzee proteins, see M.-C. King and A. C. Wilson, "Evolution at Two Levels in Humans and Chimpanzees," *Science* 188 (1975): 107–16; W. J. Bailey, K. Hayasaka, C. G. Skinner, S. Kehoe, L. C. Sieu, J. L. Slightom, and M. Goodman, "Reexamination of the African Hominoid Trichotomy with Additional Sequences from the Primate Beta-Globin Gene Cluster," *Molecular Phylogenetics and Evolution* 1 (1992): 97–135.

On differences in mitochondrial DNA among humans, chimpanzees, and gorillas, see U. Arnason, X. Xu, and A. Gullberg, "Comparison between the Complete Mitochondrial DNA Sequences of Homo and the Common Chimpanzee Based on Nonchimeric Sequences," *Journal of Molecular Evolution* 42 (1996): 145–52. On misrepresentation of the genetic similarity of humans to the apes by DNA hybridization, see Jonathan Marks, "What's Old and New in Molecular Phylogenetics," *American Journal of Physical Anthropology* 85 (1991): 207–19.

On the number of chimpanzee chromosomes, see W. J. Young, T. Merz, M. A. Ferguson-Smith, and A. W. Johnston, "Chromosome Number of the Chimpanzee, *Pan troglodytes*," *Science* 131 (1957): 1672–73.

On chromosome fusion in human lineage, see R. R. Stanyon, B. Chiarelli, and D. Romagno, "Origine del cromosoma n. 2 del cariotipo umano," *Antropologia contemporanea* 6, 3 (1983): 225–30.

On banded tips of chromosomes in chimpanzees and gorillas, see Jonathan Marks, "Hominoid Heterochromatin: Terminal C-bands as a Complex Genetic Character Linking Chimps and Gorillas," *American Journal of Physical Anthropology* 90 (1993): 237–46.

On hypothesis of two kinds of genes, see A. C. Wilson, L. R. Maxson, and V. M. Sarich, "Two Types of Molecular Evolution: Evidence from Studies of Interspecific Hybridization," *Proceedings of the National Academy of Sciences, USA* 71 (1974): 2843–47.

Blood tests for studying the similarity of human and ape proteins are described in G. H. F. Nuttall, *Blood Immunity and Blood Relationship* (Cambridge: Cambridge University Press, 1904). Nuttall's sister Zelia was an anthropologist who did important archaeological work in Mexico and was influential in the academic professionalization of the field. Small world, no?

L. M. Hussey, "The Blood of the Primates," *American Mercury* 9 (1926): 319–21.

On the similarity of hemoglobin in humans and gorillas, see Emile Zuckerkandl, "Perspectives in Molecular Anthropology," in *Classification and Human Evolution*, ed. S. L. Washburn (Chicago: Aldine, 1963), 243–72.

George Gaylord Simpson comments on the difference between gorillas and humans in "Organisms and Molecules in Evolution," *Science* 146 (1964): 1535–38.

Richard Dawkins, "Meet My Cousin, the Chimpanzee," *New Scientist*, June 5, 1993, 36–38.

On proposed classification by descent only, see W. Hennig, "Phylogenetic Systematics," *Annual Review of Entomology* 10 (1965): 97–116; K. de Queiroz and Jacques Gauthier, "Phylogenetic Taxonomy," *Annual Review of Ecology and Systematics* 23 (1992): 449–80.

On biblical dietary prohibitions, see Mary Douglas, *Purity and Danger: An Analysis of Concepts of Pollution and Taboo* (New York: Praeger, 1966).

On Linnaeus's views on breast-feeding, see Londa L. Schiebinger, *Nature's Body: Gender in the Making of Modern Science* (Boston: Beacon Press, 1993).

CHAPTER 3

This chapter is based in part on Jonathan Marks, "Science and Race," *American Behavioral Scientist* 40 (1996): 123–33, and a few other previously published essays.

On seventeenth-century perceptions of God's plan in nature, see Ivan Hannaford, *Race: The History of an Idea in the West* (Washington, D.C.: Woodrow Wilson Center Press, 1996).

François Bernier's essay was published anonymously in the *Journal des Scavans*, April 24, 1684, and translated in Thomas Bendyshe, "The History of Anthropology," *Memoirs of the Anthropological Society of London* 1 (1865): 335–458; Flavius Josephus, *The Antiquities of the Jews*, trans. William Whiston (n.p.), bk. 1, ch. 5.

The tension between the approaches of Buffon and Linnaeus to human variation is discussed in Jonathan Marks, *Human Biodiversity: Genes, Race, and History* (New York: Aldine de Gruyter, 1995).

Earnest Hooton's views on race will be found in E. A. Hooton, "Methods of Racial Analysis," *Science* 63 (1926): 75–81, and "Plain Statements about Race," *Science* 83 (1936): 511–13. And see also Hooton, *Up from the Ape* (New York: Macmillan, 1931).

Lancelot Hogben, *Genetic Principles in Medicine and Social Science* (New York: Knopf, 1932).

On race based on ABO blood-type variants, see Jonathan Marks, "The Legacy of Serological Studies in American Physical Anthropology," *History and Philosophy of the Life Sciences* 18 (1996): 345–62.

For further discussion of the Manoilov blood test, see Jonathan Marks, *Human Biodiversity*, and id., "Blood Will Tell (Won't It?)? A Century of Molecular Discourse in Anthropological Systematics," *American Journal of Physical Anthropology* 94 (1994): 59–79. Additional information is drawn from correspondence in the archives of the American Philosophical Society.

Frederick Goodwin was quoted in all the newspapers in February–March 1992. The full statement, a very meaty text, is: "If you look, for example, at male monkeys, especially in the wild, roughly half of them survive to adulthood. The other half die by violence. That is the natural way of it for males, to knock each other off and, in fact, there are some interesting evolutionary implications of that because the same hyperaggressive monkeys who kill each other are also hypersexual, so they copulate more and therefore they reproduce more to offset the fact that half of them are dying. Now, one could say that if some of the loss of structure in this society, and particularly in the high impact inner city areas, has removed some of the civilizing evolutionary things that we have built up and that maybe it isn't just careless use of the word when people call certain areas of certain cities jungles, that we may have gone back to what might be more natural, without all of the social controls that we have imposed upon ourselves as a civilization over thousands of years in our own evolution." See P. R. Breggin, "Campaigns against Racist Federal Programs by the Center for the Study of Psychiatry and Psychology," *Journal of African American Men* 1 (1995–96): 3–22.

"A 1996 book on 'apes and the origins of human violence'": Richard Wrangham and Dale Peterson, *Demonic Males: Apes and the Origins of Human Violence* (Boston: Houghton Mifflin, 1996). For comment, see L. Gould, "Negro = Man," *American Anthropologist* 67 (1965): 1281–82. The paperback version of *Demonic Males* toned down the illustration slightly by showing Adolph Schultz's full rendering (human, gorilla, chimpanzee, orangutan) on the front, rather than just the human and gorilla; a subsequent reprint of the paperback has a tasteful abstract design instead.

CHAPTER 4

This chapter incorporates some material from Jonathan Marks, "The Limits of Our Knowledge: Abilities, Responses, and Responsibilities," *Anthropology Newsletter* 36, 4 (1997): 72.

Carleton Coon's correspondence is in the National Anthropological Archives at the Smithsonian Institution. S. L. Washburn, "The Study of Race," *American Anthropologist* 65 (1963): 521–31; Ashley Montagu, "What Is Remarkable about Varieties of Man Is Likenesses, Not Differences," *Current Anthropology* 4 (1963): 361–64.

On Samuel Morton, see S. J. Gould, "Morton's Ranking of Races by Cranial Capacity," *Science* 200 (1978): 503–9; id., *The Mismeasure of Man* (New York: Norton, 1981); J. S. Michael, "A New Look at Morton's Craniological Research," *Current Anthropology* 29 (1988): 348–54.

J. P. Rushton, "Differences in Brain Size," *Nature* 358 (1992): 532; Colin Groves, "Genes, Genitals, and Genius: The Evolutionary Ecology of Race," in *Human Biology: An Integrative Science. Proceedings of the Fourth Conference of the Australasian Society for Human Biology,* ed. P. O'Higgins, 419–32 (Perth: University of Western Australia, Centre for Human Biology, 1990); L. Lieberman, "How 'Caucasoids' Got Such Big Crania and Why They Shrank: From Morton to Rushton," *Current Anthropology* 42 (2001): 69–95.

R. C. Lewontin, "The Apportionment of Human Diversity," *Evolutionary Biology* 6 (1972): 381–98.

R. L. Cann, M. Stoneking, and A. C. Wilson, "Mitochondrial DNA and Human Evolution," *Nature* 325 (1987): 31–36.

S. Tishkoff, E. Dietzsch, W. Speed, A. J. Pakstis, J. Kidd, K. Cheung, B. Bonne-Tamir, A. Santachiara-Benerecetti, P. Moral, M. Krings et al., "Global Patterns of Linkage Disequilibrium at the CD4 Locus and Modern Human Origins," *Science* 271 (1996): 1380–87.

Amos Deinard, "The Evolutionary Genetics of the Chimpanzees" (Ph.D. thesis, Department of Anthropology, Yale University, 1997).

On studies of immigrant groups, see H. L. Shapiro, *Migration and Environment* (New York: Oxford University Press, 1939); B. Kaplan, "Environment and Human Plasticity," *American Anthropologist* 56 (1954): 781–99.

On Neandertal mitochondrial DNA, see M. Krings, A. Stone, R. W. Schmitz, H. Krainitzki, M. Stoneking, and S. Paabo, "Neandertal DNA Sequences and the Origin of Modern Humans," *Cell* 90 (1997): 19–30; M. Krings, H. Geisert, R. W. Schmitz, H. Krainitzki, and S. Paabo, "DNA Sequence of the Mitochondrial Hypervariable Region II from the Neandertal Type Specimen," *Proceedings of the National Academy of Sciences, USA* 96 (1999): 5581–85; I. V. Ovchinnikov, A. Gotherstrom, G. P. Romanova, V. M. Kharitonov, K. Liden, and W. Goodwin, "Molecular Analysis of Neanderthal DNA from the Northern Caucasus," *Nature* 404 (2000): 490–93.

M. Nei and M. Roychoudhury, "Evolutionary Relationships of Human Populations on a Global Scale," *Molecular Biology and Evolution* 10 (1993): 927–43.

On mutations in mitochondrial DNA, see C. Vilà, P. Savolainen, J. E. Maldonado, I. R. Amorim, J. E. Rice, R. L. Honeycutt, K. A. Crandall, J. Lundeberg, and R. K. Wayne, "Multiple and Ancient Origins of the Domestic Dog," *Science* 276 (1996): 1687–89; T. J. Parsons, D. S. Muniec, K. Sullivan, N. Woodyatt, R. Alliston-Greiner, M. Wilson, D. Berry, K. Holland, V. Weeden, P. Gill et al., "A High Observed Substitution Rate in the Human Mitochondrial DNA Control Region," *Nature Genetics* 15 (1997): 363–67.

On species status of Neandertals: J. H. Relethford, "Ancient DNA and the Origin of Modern Humans," *Proceedings of the National Academy of Sciences, USA* 98 (2001): 390–91.

CHAPTER 5

This chapter is partly based on Jonathan Marks, "Skepticism about Behavioral Genetics," in *Exploring Public Policy Issues in Genetics,* ed. M. S. Frankel (Washington, D.C.: American Association for the Advancement of Science, 1997), 159–72.

Midcentury study of fruit flies: A. J. Bateman, "Intra-sexual Selection in Drosophila," *Heredity* 2 (1948): 349–68.

"Can Social Behavior of Man be Glimpsed in a Lowly Worm?" *New York Times,* September 8, 1998; Jonathan Marks, "Science Wars Revisited," *Anthropology Newsletter,* November 1998, 5.

On Lesch-Nyhan syndrome: T. Beardsley, "Crime and Punishment," *Scientific American,* December 1995, 22.

On MAOA: H. G. Brunner, M. Nelen, X. O. Breakefield, H. H. Ropers, and B. A. van Oost, "Abnormal Behavior Associated with a Point Mutation in the Structural Gene for Monoamine Oxidase A," *Science* 262 (1993): 578–80; V. Morell, "Evidence Found for a Possible 'aggression gene,'" *Science* 260 (1993): 1722–23. H. G. Brunner, "MAOA Deficiency and Abnormal Behaviour: Perspectives on an Association," in *Genetics of Criminal and Antisocial Behaviour,* Ciba Foundation Symposium 194 (New York: Wiley, 1995), 155–67.

E. O. Wilson, *Consilience: The Unity of Knowledge* (New York: Knopf, 1998).

Richard J. Herrnstein and Charles Murray, *The Bell Curve: Intelligence and Class Structure in American Life* (New York: Free Press, 1994).

On left-handedness, see Robert Hertz, *Death and the Right Hand,* trans. Rodney and Claudia Needham, with an introduction by E. E. Evans-Pritchard (Glencoe, Ill.: Free Press, 1960).

"In a chart listing behaviors . . .": R. Plomin, Michael J. Owen, and Peter McGuffin, "The Genetic Basis of Complex Behaviors," *Science* 264 (1994): 1733–39.

"Genes Are Tied to Homosexuality and Schizophrenia": N. Angier, "Gene Hunters Pursue Elusive and Complex Traits of Mind," *New York Times,* October 31, 1995.

On homosexuality: S. Levay, "A Difference in Hypothalamic Structure between Heterosexual and Homosexual Men," *Science* 253 (1991): 1034–37; N. Angier, "Zone of Men's Brain Linked to Sexual Orientation," *New York Times,* August 30, 1991, A1; D. Gelman, D. Foote, T. Barrett, and M. Talbot, "Born or Bred?" *Newsweek,* February 24, 1992, 46–53; W. Byne and Bruce Parsons, "Human Sexual Orientation: The Biologic Theories Reappraised," *Archives of General Psychology* 50 (March 1993): 228–38; D. Hamer, S. Hu, V. L. Magnuson, N. Hu, and A. M. L. Pattatucci, "A Linkage between DNA Markers on the X Chromosome and Male Sexual Orientation," *Science* 261 (1993): 321–27; N. Risch, E.

Squires-Wheeler, and B. Keats, "Male Sexual Orientation and Genetic Evidence," *Science* 262 (1993): 2063–65.

On eugenics: Madison Grant, *The Passing of the Great Race* (New York: Scribner's, 1916); F. A. Woods, review of id., *Science* 48, 1243 (1918): 419–20; Stefan Kühl, *The Nazi Connection: Eugenics, American Racism, and German National Socialism* (New York: Oxford University Press, 1994); C. B. Davenport, *Heredity in Relation to Eugenics* (New York: Holt, 1911); E. W. Sinnott, and L. C. Dunn, *Principles of Genetics* (New York: McGraw-Hill, 1925).

On crime: E. A. Hooton, *The American Criminal: An Anthropological Study*, vol. 1: *The Native White Criminal of Native Parentage* (Cambridge, Mass.: Harvard University Press, 1939); id., *Crime and the Man* (New York: Macmillan, 1940); R. K. Merton and M. F. Ashley-Montagu, "Crime and the Anthropologist," *American Anthropologist* 42 (1940): 384–408; P. A. Jacobs, M. Brunton, M. M. Melville, R. P. Brittain, and W. F. McClemont, "Aggressive Behaviour, Mental Sub-normality, and the XYY Male," *Nature* 208 (1965): 1351–52; B. Glass, "Science: Endless Horizons or Golden Age?" *Science* 171 (1971): 23–29; E. B. Hook, "Behavioral Implications of the Human XYY Genotype," *Science* 179 (1973): 139–50; H. A. Witkin, S. A. Mednick, F. Schulsinger, E. Bakkestrom, K. O. Christiansen, D. R. Goodenough, K. Hirschhorn, C. Lundsteen, D. R. Owen, J. Philip, D. R. Rubin, and M. Stocking, "Criminality in XYY and XXY Men," *Science* 193 (1976): 547–55; Patricia A. Jacobs, "The William Allan Memorial Award Address: Human Population Cytogenetics—The First Twenty-five Years," *American Journal of Human Genetics* 34 (1982): 689–98; G. Allen, "The Biological Basis of Crime: An Historical and Methodological Study," *Historical Studies in the Physical and Biological Sciences* 31 (2001): 183–222.

J. Q. Wilson and R. J. Herrnstein, *Crime and Human Nature* (New York: Simon & Schuster, 1985).

E. A. Hooton, *Up from the Ape* (New York: Macmillan, 1931), 491.

CHAPTER 6

This chapter is partly based on Jonathan Marks, review of *Taboo: Why Black Athletes Dominate Sports and Why We Are Afraid to Talk About It*, by Jon Entine (New York: PublicAffairs, 2000), *Human Biology* 72 (2000): 1074–78, and "Folk Heredity," in *Race and Intelligence: Separating Science from Myth*, ed. J. Fish (New York: Lawrence Erlbaum, 2001).

On taxonomism: W. C. Boyd, "Genetics and the Human Race," *Science* 140 (1963): 1057–65; A. M. Bowcock, J. R. Kidd, J. L. Mountain, J. M. Hebert, L. Carotenuto, K. K. Kidd, and L. L. Cavalli-Sforza, "Drift, Admixture, and Selection in Human Evolution: A Study with DNA Polymorphisms," *Proceedings of the National Academy of Sciences, USA* 88 (1991): 839–43; L. Cavalli-Sforza, "The Genetics of Human Populations," *Scientific American* 231, 3 (1974): 81–89; M. Nei and A. Roychoudhury, "Genic Variation within and between the Three Major

Races of Man, Caucasoids, Negroids, and Mongoloids," *American Journal of Human Genetics* 26 (1974): 421–43; D. M. Schneider, *American Kinship,* 2d ed. (Chicago: University of Chicago Press, 1980).

On racism: Jared M. Diamond, *The Third Chimpanzee: The Evolution and Future of the Human Animal* (New York: HarperCollins, 1992); Entine, *Taboo;* John M. Hoberman, *Darwin's Athletes: How Sport Has Damaged Black America and Preserved the Myth of Race* (Boston: Houghton Mifflin, 1997).

On hereditarianism: James Watson is quoted in L. Jaroff, "The Gene Hunt," *Time,* March 20, 1989, 62–67. More recently, Watson caused a row at a genetics lecture in Berkeley on October 12, 2000, when he associated skin color and hair density with sexual appetite and body build with personality. See Tom Abate, "Nobel Winner's Theories Raise Uproar in Berkeley: Geneticist's Views Strike Many as Racist, Sexist," *San Francisco Chronicle,* November 13, 2000. Other references for this section: D. E. Koshland, Jr., "Frontiers in Neuroscience," *Science* 262 (1993): 635; Lawrence Wright, *Twins: And What They Tell Us About Who We Are* (New York: Wiley, 1997); S. Begley, "All about Twins," *Newsweek,* November 23, 1987, 58–69; Constance Holden, "Identical Twins Reared Apart," *Science* 207 (March 21, 1980): 1323–27.

On essentialism: D. Singh, "Body Shape and Women's Attractiveness: The Critical Role of Waist-to-Hip Ratio," *Human Nature* 4 (1993): 297–321; D. W. Yu and G. H. Shepard, "Is Beauty in the Eye of the Beholder?" *Nature* 326 (1998): 391–92; David Buss, *The Evolution of Desire* (New York: Basic Books, 1994); M. Small, "Are We Losers? Putting a Mating Theory to the Test," *New York Times,* March 30, 1999.

H. E. Fisher, *The Anatomy of Love* (New York: Norton, 1992); Alice Eagly, "The Science and Politics of Comparing Men and Women," *American Psychologist* 50 (1995): 145–58.

F. B. M. de Waal, "The End of Nature versus Nurture," *Scientific American* 281, 6 (1999): 94–99.

CHAPTER 7

This chapter draws on Jonathan Marks, review of *Demonic Males: Apes and the Origins of Human Violence,* by Richard Wrangham and Dale Peterson (Boston: Houghton Mifflin, 1996), *Human Biology* 70 (1998): 143–46.

S. Sperling, "Baboons with Briefcases: Feminism, Functionalism, and Sociobiology in the Evolution of Primate Gender," *Signs* 17 (1991): 1–27.

Jane Goodall, *In the Shadow of Man* (Boston: Houghton Mifflin, 1971); id., *The Chimpanzees of Gombe: Patterns of Behavior* (Cambridge, Mass.: Harvard University Press, 1986); id., *Through a Window: My Thirty Years with the Chimpanzees of Gombe* (Boston: Houghton Mifflin, 1990); Dale Peterson and Jane Goodall, *Visions of Caliban: On Chimpanzees and People* (Boston: Houghton Mifflin, 1993).

Man the Hunter, ed. Richard B. Lee and Irven DeVore (Chicago: Aldine, 1969); R. B. Lee, "Art, Science, or Politics? The Crisis in Hunter-Gatherer Studies," *American Anthropologist* 94 (1992): 31–54.

Patrick Tierney, *Darkness in El Dorado: How Scientists and Journalists Devastated the Amazon* (New York: Norton, 2000).

F. B. M. de Waal, *Good Natured: The Origins of Right and Wrong in Humans and Other Animals* (Cambridge, Mass.: Harvard University Press, 1996).

Craig Stanford, *The Hunting Apes: Meat Eating and the Origins of Human Behavior* (Princeton: Princeton University Press, 1999).

CHAPTER 8

This chapter draws on Jonathan Marks, review of *The Great Ape Project: Equality beyond Humanity,* ed. Paola Cavalieri and Peter Singer (1993; 1st U.S. ed., New York: St. Martin's Press, 1994), *Human Biology* 66 (1994): 1113–17, and on some brief essays by the author in *Anthropology Newsletter.*

Franz Boas, *The Mind of Primitive Man,* rev. ed. (New York: Macmillan, 1938).

Joel Wallman, *Aping Language* (New York: Cambridge University Press, 1992).

A. Whiten, J. Goodall, W. C. McGrew, T. Nishida, V. Reynolds, Y. Sugiyama, C. E. G. Tutin, R. W. Wrangham, and C. Boesch, "Cultures in Chimpanzees," *Nature* 399 (1999): 682–85.

A. Whiten and C. Boesch, "The Cultures of Chimpanzees," *Scientific American,* January 2001, 61–67.

"Chimpanzees with Culture" (editorial), *New York Times,* July 7, 1999.

B. Meggers, "Mate Copying and Cultural Inheritance," *Trends in Ecology and Evolution* 13 (1998): 240.

R. Tuttle, "On Culture and Traditional Chimpanzees," *Current Anthropology* 42 (2001): 407–8.

The Great Ape Project: Equality beyond Humanity, ed. Paola Cavalieri and Peter Singer (London: Fourth Estate, 1993).

Daniel Povinelli, *Folk Physics for Apes: The Chimpanzee's Theory of How the World Works* (New York: Oxford University Press, 2000).

Section 85 of the Animal Welfare Act passed by the New Zealand Parliament in 1999 places "restrictions on [the] use of non-human hominids." Its target specifically is scientific research (which comprises a minuscule proportion of the problems affecting and threatening the apes). Although supporters of ape rights hailed it as a victory (for example, in the *New York Times,* August 12, 2001), it does not use the words "rights" or "apes." In fact, the phrase "non-human hominids" would seem to allow a loophole to any researcher who cared to use the traditional classification, which restricts "hominids" to humans, and calls the apes "pongids." At face value, the law might be seen simply as placing limits on experimentation upon australopithecines.

CHAPTER 9

J. Walsh and J. Marks, "Sequencing the Human Genome," *Nature* 322 (1986): 590.

L. L. Cavalli-Sforza and A. W. F. Edwards, "Analysis of Human Evolution," in *Genetics Today: Proceedings of the XI International Congress of Genetics*, ed. S. J. Geerts (Oxford: Pergamon, 1965), 923–33; M. Nei and A. Roychoudhury, "Genic Variation within and between the Three Major Races of Man, Caucasoids, Negroids, and Mongoloids," *American Journal of Human Genetics* 26 (1974): 421–43.

"Cavalli-Sforza was prepared to acknowledge . . . mistaken": L. L. Cavalli-Sforza, A, Piazza, P. Menozzi, and J. Mountain, "Reconstruction of Human Evolution: Bringing Together Genetic, Archaeological, and Linguistic Data," *Proceedings of the National Academy of Sciences, USA* 85 (1988): 6002–6.

The Human Genome Diversity Project: L. L. Cavalli-Sforza, A. C. Wilson, C. R. Cantor, R. M. Cook-Deegan, and M.-C. King, "Call for a Worldwide Survey of Human Genetic Diversity: A Vanishing Opportunity for the Human Genome Project," *Genomics* 11 (1991): 490–91; L. Roberts, "A Genetic Survey of Vanishing Peoples," *Science* 252 (1991): 1614–17; J. M. Diamond, "A Way to World Knowledge," *Nature* 352 (1991): 567; L. Roberts, "Genetic Survey Gains Momentum," *Science* 254 (1991): 517; id., "Genome Diversity Project: Anthropologists Climb (Gingerly) On Board," *Science* 258 (1992): 1300–1301.

Criticism of the HGDP: Jonathan Marks, "The Human Genome Diversity Project: Good *For* if Not Good *As* Anthropology?" *Anthropology Newsletter* 36 (April 1995): 72; J. C. Gutin, "End of the Rainbow," *Discover,* November 1994, 70–75; S. Subramanian, "The Story in Our Genes," *Time,* January 16, 1995, 54–55; A. M. Bowcock, J. R. Kidd, J. L Mountain, J. M. Hebert, L. Carotenuto, K. K. Kidd, and L. L. Cavalli-Sforza, "Drift, Admixture, and Selection in Human Evolution: A Study with DNA Polymorphisms," *Proceedings of the National Academy of Sciences, USA* 88 (1991): 839–43; G. Taubes, "Scientists Attacked for 'patenting' Pacific Tribe," *Science* 270 (1995): 1112. And see also Carleton S. Coon, *The Story of Man* (New York: Knopf, 1954); M. F. A. Montagu, "The Concept of Race in the Human Species in the Light of Genetics," *Journal of Heredity* 32 (1941): 243–47.

National Research Council, *Evaluating Human Genetic Diversity* (Washington, D.C.: National Academy Press, 1997); E. Pennisi, "NRC OKs Long-Delayed Survey of Human Genome Diversity," *Science* 278 (1997): 568; V. Dominguez, "Misleading News Coverage Concerning HGDP," *Anthropology Newsletter,* January 1998, 18; E. Marshall, "DNA Studies Challenge the Meaning of Race," *Science* 282 (1998): 654–55.

On going beyond microphylogenetic issues, see, e.g., L. G. Carvajal-Carmona, I. D. Soto, N. Pineda, B. Ortiz, C. Daniel, C. Duque, J. Ospina-Duque, M. McCarthy, P. Montoya, V. M. Alvarez, G. Bedoya et al., "Strong Amerind/White Sex Bias and a Possible Sephardic Contribution among the Founders of a Population in Northwest Colombia," *American Journal of Human Genetics* 67 (2000): 1287–95. This paper shows that a population with a known history from Colombia has 94 percent European Y chromosomes, 90 percent Native American mitochondrial DNA, and 70 percent European blood-group markers. Clearly any statements about the relationships among contemporary populations with other histories will depend mightily upon what particular genes are being compared!

CHAPTER 10

On the subjects discussed in this chapter, see also Jonathan Marks, "Racism, Eugenics, and the Burdens of History," *Yale Journal of Ethics* 5 (1996): 12–15, 40–42.

The experimental hybridization of human and ape has been seriously proposed in the scientific literature, for example, by biologist Charles Remington, in *Experimentation with Human Beings: The Authority of the Investigator, Subject, Professions, and State in the Human Experimentation Process,* ed. Jay Katz (New York: Russell Sage Foundation, 1972), 461–64.

Paul R. Ehrlich, *Human Natures: Genes, Cultures, and the Human Prospect* (Washington, D.C.: Island Press for Shearwater Books, 2000).

D. W. Brock, "Cloning Human Beings: An Assessment of the Ethical Issues Pro and Con," in *Clones and Clones: Facts and Fantasies about Human Cloning,* ed. M. C. Nussbaum and C. R. Sunstein (New York: Norton, 1998), 141–64.

On the Native American Grave Protection and Repatriation Act and Kennewick Man, see also David Hurst Thomas, *Skull Wars: Kennewick Man, Archaeology, and the Battle for Native American Identity* (New York: Basic Books, 2000).

For James Chatters's side of the story, see J. C. Chatters, "Encounter with an Ancestor," *Anthropology Newsletter,* January 1997, 9–10; id., "Response from Chatters," ibid., May 1997, 6.

There is no primary scientific literature on the subject of the Kennewick Man. Reports include T. Egan, "Tribe Stops Study of Bones That Challenge History," *New York Times,* September 30, 1996; Douglas Preston, "The Lost Man," *New Yorker,* June 16, 1997, 70–78, 80–81; C. W. Petit, "Redisovering America," *U.S. News & World Report,* October 12, 1998, 56–64; N. Charles and J. Hannah, "Head Case: Archaeologist James Chatters Speaks for Science in a Fight over the Fate of a 9,000 Year-Old Skeleton," *People,* November 30, 1998, 181–82. Both the *Seattle Times* and the *Tri-City Herald* have on-line archives. See esp. John Stang, "Skull Found on Shore of Columbia," *Tri-City Herald,* July 29, 1996; Mike Lee, "Tri-Citians Sculpt Theoretical Look for Kennewick Man," ibid., Feb 10, 1998, and "Missing Ancient Bones Discovered," ibid, June 22, 2001.

CHAPTER 11

K. Skorecki, S. Selig, S. Blazer, R. Bradman, N. W. P. J. Bradman, M. Ismajlowiscz, and M. Hammer, "Y Chromosomes of Jewish Priests," *Nature* 385 (1997): 32; D. Grady, "Finding Genetic Traces of Jewish Priesthood," *New York Times,* January 7, 1997.

V. A. McKusick, R. Eldridge, R. A. Hostetler, and J. A. Egeland, "Dwarfism in the Amish," *Transactions of the Association of American Physicians* 77 (1964): 151–68.

M. G. Thomas, K. Skorecki, H. Ben-Ami, T. Parfitt, N. Bradman, and D. Goldstein, "Origins of Old Testament Priests," *Nature* 394 (1998): 138–39.

The only other study associating surnames and Y-chromosomes looked at forty-eight men named Sykes in three English counties, without knowing how many, if any, were known relatives. Half of them were found to have the same Y-chromosome, which the researchers assumed derived from an original Sykes in late medieval times. The other Sykeses in the group were attributed either to false paternity or to adoption over the centuries. See B. Sykes and C. Irven, "Surnames and the Y Chromosome," *American Journal of Human Genetics* 66 (2000): 1417–19.

Methodological problems with the Cohanim study are detailed in A. Zoossmann-Diskin, "Are Today's Jewish Priests Descended from the Old Ones?" *Homo* 51 (2000): 156–62.

On Jewish genetic data invested with religious authority, see D. Grady, "Who Is Aaron's Heir? Father Doesn't Always Know Best," *New York Times,* January 19, 1997; P. Herschberg, "Decoding the Priesthood," *Jerusalem Report,* May 10, 1999, 30–35.

D. M. Schneider, *American Kinship: A Cultural Account,* 2d ed. (Englewood Cliffs, N.J.: Prentice-Hall, 1980). See also K. Finkler, "The Kin in the Gene," *Current Anthropology* 42 (2001): 235–63.

M. G. Thomas, T. Parfitt, D. A. Weiss, K. Skorecki, J. Wilson, M. le Roux, N. Bradman, and D. Goldstein, "Y Chromosomes Traveling South: The Cohen Modal Haplotype and the Origins of the Lemba—the 'Black Jews' of Southern Africa," *American Journal of Human Genetics* 66 (2000): 674–86.

On creationism, see Ronald L. Numbers, *The Creationists* (1992; Berkeley and Los Angeles: University of California Press, 1993); Phillip E. Johnson, *Darwin on Trial* (Washington, D.C.: Regnery Gateway, 1991). Quotations taken from Phillip E. Johnson, *Objections Sustained: Subversive Essays on Evolution, Law, and Culture* (Downer's Grove, Ill.: InterVarsity Press, 1998), 20. See also B. Appleyard, "You Asked for It," *New Scientist,* April 22, 2000.

On social Darwinism, see Peter J. Bowler, "Social Metaphors in Evolutionary Biology, 1870–1930: The Wider Dimension of Social Darwinism," in *Biology as Society, Society as Biology: Metaphors,* ed. Sabine Maasen, Everett Mendelsohn, and Peter Weingart (Boston: Kluwer Academic Publishers, 1995), 107–26.

M. Goodman, "Epilogue: A Personal Account of the Origins of a New Paradigm," *Molecular Phylogenetics and Evolution* 5 (1996): 269–85; V. M. Sarich, "The Origin of the Hominids: An Immunological Approach," in *Perspectives on Human Evolution*, vol. 1, ed. S. L. Washburn and P. C. Jay (New York: Holt, Rinehart & Winston, 1968), 94–121.

On the Human Genome Evolution Project, see A. Gibbons, "Which of Our Genes Make Us Human?" *Science* 281 (1998): 1432–44; E. McConkey and A. Varki, "A Primate Genome Project Deserves High Priority," ibid., 289 (2000): 1295.

CHAPTER 12

This chapter is partially based on Jonathan Marks, "The Anthropology of Science, Part I: Science as a Humanities" and "The Anthropology of Science, Part II: Scientific Norms and Behaviors," *Evolutionary Anthropology* 5 (1996): 6–10, 75–80. Extensive references can be found there.

I. Parker, "Richard Dawkins's Evolution: An Irascible Don Becomes a Surprising Celebrity," *New Yorker,* September 9, 1996; Richard Dawkins, *River Out of Eden: A Darwinian View of Life* (New York: Basic Books, 1995).

Alan Sokal and Jean Bricmont, *Fashionable Nonsense: Postmodern Intellectuals' Abuse of Science* (New York: Picador USA, 1998).

Clarence Darrow, "The Eugenics Cult," *American Mercury* 8 (1926): 129–37.

"One molecular geneticist writes . . . ".: S. J. Jones in *The Cambridge Encyclopedia of Human Evolution* (New York: Cambridge University Press, 1992), 442.

N. Wade, "Genetic Code of Human Life Is Cracked by Scientists," *New York Times,* June 27, 2000.

E. Gosney and P. Popenoe, *Sterilization for Human Betterment* (New York: Macmillan, 1929).

Mary Midgley, *Science as Salvation* (New York: Routledge, 1992).

E. O. Wilson, *Consilience: The Unity of Knowledge* (New York: Knopf, 1998).

Charles Davenport, *Heredity in Relation to Eugenics* (New York: Holt, 1911).

Erwin Schrödinger, *What Is Life?* (Cambridge: Cambridge University Press, 1944).

Richard Dawkins, *River Out of Eden: A Darwinian View of Life* (New York: Basic Books, 1995), 132–33.

José Ortega y Gasset, *The Revolt of the Masses* (New York: Norton, 1932).

"As one commentator has noted": Bernard Dixon, *What Is Science For?* (London: Penguin Books, 1973), 68.

Ashley Montagu to Theodosius Dobzhansky: quoted from Susan Sperling's obituary of Montagu in the *American Anthropologist* (2000).

INDEX

Page numbers appearing in italics refer to illustrations.

Aaron (Moses's brother), 245, 246–47
abilities, 91–95. *See also* potentials/abilities
achondroplasia, 106, 264
Aesop, 103
Africans: genetic diversity among, 83–84, 245; as paraphyletic, 84, 134; as a race, 52–53, 54, 66–67, 131–33
aggression, 105, 107–8, 124; in males, 159–60, 162–63, 172–74. *See also* violence
AIDS, 265
alcoholism studies, 216–17
alienness/otherness, 141–42
Allen, Woody: *Sleeper,* 223
altruism, 250
American Mercury, 42
Americas, discovery of, 166
Amerindians. *See* Native Americans
Amish people (Pennsylvania), 247
anatomy vs. genetics, 23–29, 31, 200
Ancient Hunters (Sollas), 260
animal rights movement, 185, 188–89. *See also* apes, human rights for
Annals of Improbable Research (Harvard), 30

anthropology, 2, 206
anthropomorphism, 103
anti-intellectualism, 285
Antiquities of the Jews (Josephus), 53–54
apes: conservation of, 188–89; DNA across, 34; as endangered, 185, 186; evolution of, 9; genetic diversity among, 86–87; as models for human nature, 160–61; and primates' hands/feet, 21–22; species of, 8–9; stereotypes of, 69–71, *70,* 293. *See also* apes, human rights for; chimpanzees; human–ape genetic similarity
apes, human rights for, 180–97; apes as disabled people, 189–93; and the Bangkok Six, 193–95; and culture, 181–85; The Great Ape Project, 5–6, 185–89; and human rights violations, 192–94; legislation on, 196–97, 298; and mental capacities of apes, 195–96; and populations of humans vs. apes, 194–95; and responsibility, 192
Arendt, Hannah, 186–87
Aristotle, 49

Army Corps of Engineers, 229, 230, 232, 234

Arunta people (Australia), 169

Asians, 53, 54, 64, 66–67, 131–32

Australian aborigines, 169, 179, 240

Babbitt, Bruce, 253

baboons, 11, 24–25, 161–63

Bacon, Francis, 254, 272–73

Bangkok Six, 193–95

Basques, 133

Bateman, A. J., 102

Bateson, William, 128

Battell, Andrew, 13–14

behavioral genetics, 5, 100–127; characteristics (phenotypes) vs. hereditary states (genotypes), 100–101, 157–58; on crime, 122–26; cultural bases of, 112–13, 119, 127 (*see also* folk heredity); and diseases, gene mapping for, 105–6; and fruit flies, 101–2; and homology, 104; of homosexuality, 110–17, 125, 156–57; human behaviorial variation, scope of, 117–19; and maleness, 123–24; molecular anthropological perspective on, 122–27; on morality, 120–21; nature–vs.–nurture debate, 100, 108–10, 158; responsibility/pronouncements of human genetics, 119–22, 157; studying human behavior, 105–8; of worms vs. humans, 103–4

The Bell Curve (Herrnstein and Murray): and Cavalli-Sforza's *History and Geography of Human Genes*, 210; on cognitive ability, 91–92; genetic data lacking in, 143–44; on IQ/social status as innate, 108–9, 130, 146, 180; racist work cited in, 150; reception of, 5, 130; on social programs, 91–92, 108–9

Benedict, Ruth, 74

Bible, 246–47, 254–55

Bijan, 30

bioethics, 208–9, 212–14, 215

biological systematics, 19–20

birds and reptiles, 44–45

blacks, athletic superiority of, 80, 143–46

blood, 242–65; and ape–human genetic similarity, significance of, 261–63; and biogenetic substance, 249–51; and creationism, 255–61; essentializing of, 243–44, 245–49; and genes, shared proportion of, 250; and genetic similarity among people with similar surnames, 245–48, 301; and genetic tests for race, 244–45; as heredity, 243–44, 251; and human–chimpanzee kinship, as threatening to religion, 254–55, 257; and the Human Genome Evolution Project, 263–65; and Kennewick Man, 251–53; as a metaphor/symbol, 43–44, 127, 242–43, 249; and relatives/kinship, 243, 249–51

blood tests for race, 63–65, 81, 126–27

Blumenbach, Friedrich, 21, 61, 65

Boas, Franz: anthropological school of, 73, 74; attitude toward science, 276; on culture, 181–82; on Grant's *Passing of the Great Race,* 120; on skull size/shape, 90–91, 177; on technology, 272

Bodmer, Sir Walter, 200

Bonnichsen, Robson, 232, 233

bonobos, 8, 110–11, 174, 175

Bosman, Willem, 17–18

Bouchard, Thomas, 149–51

Boyd, William C., 132–33

The Boys from Brazil (Levin), 223

Brace, Loring, 232

Bricmont, Jean, 268

Brinton, Daniel Garrison, 236

Brunner, Han, 107–8

Bryan, William Jennings, 257, 258–59, 261

Buck v. Bell, 120

Buffon, Georges-Louis Leclerc, comte de, 59–61, 65, 68, 70

Burt, Sir Cyril, 114

Buss, David, 155

Cann, Rebecca, 83

Cantor, Charles, 202

Cavalieri, Paola: *The Great Ape Project,* 186, 189

Cavalli-Sforza, Luca, 134–35, 199–200, 201–2, 207; *The History and Geography of Human Genes,* 209–10, 211

Cave Allegory (Plato), 58

C-banding, 39, *39*

cell lines, patents on, 212–14

Chang and Eng (Siamese Twins), 224

Chatters, James, 228–29, 230, 232, 234–36

"chimpanzee," as a racial insult, 69–71, *71,* 293

chimpanzee-human hybrids, 219–21

chimpanzees: aggression in, 159–60, 172–73; in captivity, 185; culture attributed to, 181, 182, 184–85; evolution of, 9–10; Goodall's Gombe studies of, 163–65; humans' evolutionary divergence from, 11; humans' shared ancestry with, as threatening to religion, 254–55, 257; humans' similarities to, 7–8, 245 (*see also* human–ape genetic similarity); language attributed to, 182–84; as model for evolution of human behavior, 163–65; as model for "natural man," 172–75, 176; plenitude/range of, 8; sexuality in, 110, 188; Taï and Mahale studies of, 165; tools used by, 22, 182

Christianity, 140

chromosomes: function and structure of, 36–39, *38–39;* vs. genes, 36; number of, in humans, 30–31, 37

citizenship and equal rights, 186–87

civilization, 76–77, 181, 260

civil rights movement, 75

classification: Buffon on, 59–61; cultural bases of, 48–50, 68 (*see also* folk heredity); Hebrew, of animals, 46–47; by kinship, 47–48; Linnaean, 19–20, 46, 59–61, 68 (see also *Systema naturae*); by race (*see* race)

Clinton, Bill, 271

cloning, 223–25

coelacanths, 45

cognitive ability, 91–92

Cohanim, Y chromosomes of, 245–46, 247–49, 251

coincidences, 150

Collins, Francis, 271

Columbia University Medical School, 74

Cook, Captain James, 56

Cook-Deegan, Robert, 202

Coon, Carleton Stevens, 75, 208, 209; *The Origin of Races,* 76–77

creationism, 191–92, 255–61, 280–81, 285–86

crime, 122–26, 139

cultural diversity, 87–88

cultural relativism, 227, 272–74

culture, 72; and apes' rights, 181–85; definition of, 181–82; traits of, 88–89

cytogenetics, 37

daffodils, 28–29

Darkness in El Dorado (Tierney), 287

Darrow, Clarence, 258–59, 269, 277

Darwin, Charles, 59, 104, 130, 269, 276; *The Descent of Man,* 261

Darwinism, 222, 236, 254. *See also* evolution

Davenport, Charles, 65, 120, 121, 148, 268–69, 270, 277

Dawkins, Richard, 44, 186, 266, 282–83

Deinard, Amos, 86–87

demonism. *See* aggression

Depression, 269

descent and identity. *See* identity and descent; Kennewick Man

The Descent of Man (Darwin), 261

de Waal, Frans, 158, 174

Diamond, Jared, 177, 178, 186, 204

Discourse on the Origin of Inequality (Rousseau), 168

diseases, gene mapping for, 41, 105–6

DNA: across apes, 34; and C-banding, 39, *39;* and chromosomal function/structure, 36–39, *38–39;* and diseases, mapping for, 41, 105–6; and genes, vs. chromosomes, 36; and genes coded for biochemical minutiae vs. bodies, 41; genic vs. nongenic, 33; hybridization of, 34–35; as measure of genetic similarity, 10, 23, 24–29, 32–33; mitochondrial, 33–34, 83, 85–86, 99; mutations to, 32–33, 34, 99; Neandertal, 96–97, 98–99; nuclear, 84–87; percentages of genetic similarity, interpreting, 29–31, 34–36; proteins in, 32–33; recombination of, 84–85; and RNA, 129; XYY research, 123–25

Dobzhansky, Theodosius, 75, 81, 287–88; *Genetics and the Origin of Species,* 74

Dolly (cloned sheep), 223
Doyle, Sir Arthur Conan, 89
Durkheim, Emile, 276; *The Elementary Forms of the Religious Life*, 169
dwarfism, 106

Eagly, Alice, 156
Ehrlich, Paul, 224
The Elementary Forms of the Religious Life (Durkheim), 169
Ellis–van Creveld Syndrome, 247
Enlightenment, 186–87
ESP, 151–52
essentialism, 58, 112, 153–57, 243–44, 245–49. *See also* human nature
eugenics, 120, 124–25, 268–71, 277
Europeans, as a race, 52, 53, 54, 55, 67, 131–32
evolution: vs. creationism, 191–92, 255–56; Darwinian theory of, 222, 236, 254; descent, patterns of, 58–59; macromutational view of, 264; and meaninglessness of life, 281; political interpretations of (*see* social Darwinism); social concerns not addressed by, 280; Victorian views of, 189–90
evolutionary psychology, 153–54
Exodus, 246–47

Falk, K. George, 65
family, 135–36. *See also* kinship
Feldman, Marcus, 210
Fischer, Eugen, 270
Fison, Lorimer, 169
folk heredity, 128–31, 146–58: essentialism, 153–57; and genetics, 129, 130–31; hereditarianism, 146–53; heredity vs. inheritance, 129–30; Law of Independent Assortment, 129; Law of Segregation, 128–29; racism, 139–46; and responsibility, 157–58; social Darwinism, 130–31, 259–61; taxonomism, 131–38
folk knowledge vs. science, 2, 5, 93
French Revolution, 277
Freud, Sigmund, 276
Friedlaender, Jonathan, 207
fruit flies, 28, 101–2

Galileo, 52
genes vs. chromosomes, 36
genetics: vs. anatomy, 23–29, 31, 200; credibility of, 147; human variation, as based on, 88–91; race, as based on, 81–82 (*see also* race); and racism/xenophobia, 2; as social panacea, 285–86. *See also* behavioral genetics; DNA; folk heredity; human–ape genetic similarity; molecular anthropology
Genetics and the Origin of Species (Dobzhansky), 74
genocide, 141–43
Genomics, 203–4
genotypes vs. phenotypes, 100–101, 157–58
gibbons, 8
Gill, George, 232
Gillen, F. J., 169
Gish, Duane, 256
Glass, Bentley, 124
God's bounty vs. parsimony, 52
Goodall, Jane, 22, 163–64, 192
Goodman, Morris, 10, 261–62
Goodwin, Frederick, 70, 293
gorillas, 8, 9–10, 14–15, 42–43
Gould, Stephen Jay, 79
Grant, Madison: *The Passing of the Great Race*, 119–20, 268–69
The Great Ape Project, 5–6, 185–89
The Great Ape Project (Cavalieri and Singer), 186, 189
great chain of being, 69
Great Genome Diversity Project, 5–6
Greely, Henry, 207
Groves, Colin, 79–80
Guinier, Lani, 69
Gypsies, 140

Hagahai people (New Guinea), 212–13
Hamer, Dean, 115–17, 125
Harry, Debra, 222–23
Haynes, Vance, 232
head size/shape, 78–80, 89–91, 177
Hebrew classification of animals, 46–47
hereditarianism, 146–53

heredity: blood as, 243–44, 251; cultural ideas about (*see* folk heredity); scientific study of (*see* genetics)

heritability, 146–47

Herodotus, 52, 166

Herrnstein, Richard, 74, 125, 149. See also *The Bell Curve*

HGDP. *See* Human Genome Diversity Project

The History and Geography of Human Genes (Cavalli-Sforza), 209–10, 211

Hitler, Adolf, 224, 268–69, 270

Hobbes, Thomas: *Leviathan*, 166–67, 173, 176, 179

Hoberman, John, 145

Hogben, Lancelot, 62–63

Holden, Constance, 152–53

Holocaust, 142

Homo erectus, 179

homology, 104

Homo sapiens monstrosus, 56

homosexuality, 110–17, 125, 156–57, 175

Hooton, Earnest: background of, 61–62; on blood test for race, 63–65, 126–27; on criminals, 123; on primary vs. secondary races, 62; on skull form as determiner of race, 236–37; and Washburn, 73, 74–75

Howitt, A. L., 169

Hrdlička, Aleš, 61

Hulse, Frederick, 91

human–ape genetic similarity, 13–31; chromosomal, 37–39, *38–39*, 245; and classification as a cultural act, 48–50; genetics vs. anatomy, 23–29, 31; in hemoglobin, 33, 42–44; historical/mythological accounts of, 13 19, *16, 18;* implications of, 4, 41–42; Linnaeus on, 21; meaning of, 5, 261–63; and molecular anthropology, central fallacy of, 41–48; and phylogenetic relatives, 44–48, *46;* vs. visible differences, 40–41. *See also* DNA

human-chimpanzee hybrids, 219–21

Human Genome Diversity Project (HGDP), 199–218; and alcoholism studies, 216–17; and bioethics, 208–9, 212–14, 215; biomedical benefits of, 207; and consent vs. coercion, 208–9, 215; death throes of, 214–16; and ethnic populations vs. genetic groupings, 202–4, 299–300; funding for, 212; and mitochondrial Eve, 201; and normality, 198–99; objections to, 205, 206–9, 217–18; proposal/support for, 199–200, 202–4, 205–6; and race, 200–202, *201*, 209–12, 216; and research priorities, 216–17; and vanishing populations, 204–5, 206

Human Genome Evolution Project, 263–65

Human Genome Project, 198–99, 213, 263, 271

humanism, 284, 285, 286, 288

humanities vs. science, 1

human nature, 159–79; apes vs. humans, 160–61; and cultural projections, 161–65; and female passivity, 162–63; head shape as, 177; and homosexuality, 156–57; and male aggression/demonism, 159–60, 162–63, 172–74; menstruation as, 178; natural man, without culture, 165–75, 176–77, 178–79; and natural sex, 175–76; polygyny vs. monogamy as, 177–78; and waist-to-hip ratio of women, 154–55

human rights, 269–70. *See also* apes, human rights for

humans: chimpanzees' shared ancestry with, as threatening to religion, 254–55, 257; differences among, 5 (*see also* race); evolutionary divergence from apes, 9–12; fossils of, 10; mental distinctiveness of, 22; orangutans' genetic similarity to, 26–28; plenitude/range of, 8; teeth of, 23. *See also* human–ape genetic similarity; human variation, meaning of

human variation, meaning of, 72–99; behaviorial, 117–19 (*see also* behavioral genetics); in cognitive ability, 91–92; genetic basis for, 88–91; in head size/shape, 89–91; among immigrants, 90–91, 117; Neandertals vs. modern humans, 95–99; and potentials/abilities, 91–95; skeletal, 72–73. *See also* race

hunter-gatherers, 169–70

hunting, 162
Huxley, Thomas, 161, 191–92, 262

identity and descent, 219–25, 239–41; and cloning, 223–25; and human-chimpanzee hybrids, 219–21; and kinship, 221–22; who and what we are, questions of, 221–23. *See also* Kennewick Man
IgNobel prize (Harvard), 30
immigration, 90–91, 117, 119–20, 121, 269
inbreeding, measures of, 247
Indians,52
indigenous peoples, 206, 217. *See also* Kennewick Man
Indonesia, 193–94
Inherit the Wind, 257, 258
In His Image (Rorvick), 223
Institute for Creation Research, 256
intermarriage, laws against, 68
IQ, 146, 149. See also *The Bell Curve*
isolationism, 273
isonymy, 247

Jack and Oskar (twins), 152–53
Jacobs, Patricia, 123–24
Jantz, Richard, 232
Jefferson, Thomas, 284
Jenkins, Carol, 212–13
Jensen, Arthur, 149, 210
Jews, 140, 245–46, 247–49, 251
Jim twins, 149–51
Johnson, Phillip, 257, 261
Josephus, Flavius: *Antiquities of the Jews,* 53–54

Kaestle, Frederika, 225–26, 229, 252
Kafka, Franz: "Report to an Academy," 40
Kamileroi people (Australia), 169
Keith, Sir Arthur, 73
Kellogg, Vernon, 261
Kennewick Man, 225–40; DNA tests on, 251–53; and human rights/respect issues, 225–28, 252; hype about, 228–37; and race, 233–36; reasons for returning him to Native Americans, 237–39
King, Mary-Claire, 32–33, 202

Kingsley, Charles, 69–70
kinship, 47–48, 221–22, 243, 249–51
Koshland, Daniel, 148
!Kung San people (Kalahari Desert), 169–71, 172, 179, 204
Kurnai people (Australia), 169

language, 182–84
Lapps, as a race, 53, 55
Latour, Bruno, 268
Law of Segregation, 128–29
Leakey, Louis, 163–64
Lee, Richard, 171
left-handedness, 109–10
Lemba people (South Africa), 248, 251
Lesch–Nyhan syndrome, 107, 108
LeVay, Simon, 113–14
Leviathan (Hobbes), 166–67, 173, 176, 179
Levin, Ira: *The Boys from Brazil,* 223
Lewontin, Richard, 81–82
Linnaeus, Carl: and Buffon, 60; career/reputation of, 20; on human–ape genetic similarity, 21; influence of, 59; life/personality of, 20; on personality and cultural traits, 88–89; on race, 55–61, 57 (table); scientific classification by, 19–20, 49–50, 59, 68; *Systema naturae,* 20–21, 49–50, 55–56, 57 (table)
Locke, John: *Second Treatise on Civil Government,* 167–68
Lucy, 240
Luzia, 240

maleness, 123–24
Malinowski, Bronislaw, 73
Malthus, Thomas, 194
mammals, 49–50
Mandeville, Sir John: *Travels of Sir John Mandeville,* 166
Manoilov, E. O., 64–65, 126–27
Manouvrier, Léonce, 61
"Man the Hunter" (1966), 170
MAOA (monoamine oxidase A) deficiency, 107–8
Marx, Karl, 276
Maslack, Fred, 244

Matthew, William Diller, 260
Mead, Margaret, 74, 287
Medawar, Peter, 267, 273–74
Mencken, H. L., 42, 246
Mendel, Gregor, 101, 128–29, 269, 279
Mengele, Josef, 252, 270
Meninick, Jerry, 229
menstruation, 178
metric system, 277
Midgley, Mary, 274
miscegenation laws, 68
mitochondrial Eve, 34, 83, 201
mitochondrion, 33–34
Mo-Cell, 214
molecular anthropology, 7–31; Central Fallacy of, 41–48; definition of, 1–2, 3; and DNA as a measure of genetic similarity, 10, 23, 24–29; genetics vs. anatomy, 23–29, 31; and human–ape evolutionary divergence, 9–12; and humanistic science, 286–88; as mediator, 3, 6; and monsters, stories of, 13 –15; and otherness of humans, 21–23; percentages of genetic similarity, interpreting, 29–31; and primates' hands/feet, 21–22; and proteins as measure of genetic similarity, 10–11, 23; and *Ramapithecus,* 11–12; and scientific classification, 19–20; value of, 3–4. *See also* DNA; human–ape genetic similarity
monoamine oxidase A. *See* MAOA deficiency
Monod, Jacques, 281
monogamy, 177–78
Monroe, Marilyn, 154
monsters, stories of, 13 –15
Montagu, Ashley, 73–74, 75, 77, 210, 287–88
Moore, John, 214
morality, 120–21
moral relativism, 273
Morgan, Lewis Henry, 169
Morgan, Thomas Hunt, 65, 128
Morton, Samuel George, 78–79
Muller-Hill, Benno, 214–15
Murray, Charles. See *The Bell Curve*
myth, 272–73, 280–84

NAGPRA (Native American Graves Protection and Repatriation Act), 206, 227–28, 229, 230, 232–33, 240
National Academy of Sciences, 215
National Institutes of Health (NIH), 122, 213
National Research Council, 215, 216
Native American Graves Protection and Repatriation Act. *See* NAGPRA
Native Americans: artifacts/skeletal remains reclaimed by, 206; and genetic tests for proof of ancestry, 244–45; as "natural man," 166–69, 179; origins of (*see* Kennewick Man); as a race, 53, 64
naturalistic fallacy, 161
natural selection, 130
Nature, 197
nature-vs.-nurture debate, 100, 108–10, 158
Navajo people, 216–17
Nazis, 75, 140, 270, 271, 277
Neandertals, vs. modern people, 95–99
Near East peoples, as a race, 52
Nebraska Man, 256
Nei, Masatoshi, 134–35
New Scientist, 196–97
Newton, Isaac, 20, 52
New World, discovery of, 166
New Yorker, 233–34
New York Times, 231–32
New York Zoological Society, 119, 268–69
New Zealand, ape-rights legislation in, 196–97, 298
NIH (National Institutes of Health), 122, 213
Noah's Sons model of racial origins, 53–54, 67, 97–98, 211–12
nonintervention, 273
Nuremberg Code, 208–9
Nuttall, George, 42
Nuttall, Zelia, 292

Onandaga Council of Chiefs, 207
On the Town, 89
orangutans: baboons' genetic similarity to, 24–25; and the Bangkok Six, 193–95; humans' evolutionary divergence from,

orangutans (*continued*)
11; humans' genetic similarity to, 26–28;
humans' similarity to, 8–9
The Origin of Races (Coon), 76–77
Ortega y Gasset, José, 285
Overton, William R., 256
Owsley, Douglas, 232

paraphyletic categories, 45–46, 50
Parche, Günther, 148
The Passing of the Great Race (Grant), 119–
20, 268–69
Pearl, Raymond, 270
Pearson, Karl, 260
Pecos Pueblo skulls, 236–37
pedophilia, 175
Pennant, Thomas, 21, 60
percentages of genetic similarity, interpret-
ing, 29–31
personality traits, 88–89
Peterson, Dale, 159, 173, 174
phenotypes vs. genotypes, 100–101, 157–58
physical anthropology, 61–69, 73–74, 140
Pilbeam, David, 11–12
Piltdown Man, 256
Pioneer Fund, 149–50
Plato, 58
Pliny (the Elder), 166
polygenists, 53
polygyny, 177–78
Popper, Karl, 267
populations, 74–75, 77, 80–82, 132–34
Porter, Cole, 275
potentials/abilities, 91–95, 180–81
poverty, 122
Povinelli, Daniel, 196
prejudice, 69–71, *71*, 94–95, 293
Preston, Douglas, 233–34
primates, hands/feet of, 21–22
proteins, 10–11, 23, 32
psychic connections, 151
Purchas, Samuel, 13–14
Putnam, Carleton, 77
Pygmie, 17–18, *18*
pyramids, 234

Quadrupedia (four-legged), 49

race, 51–71, 75–88; athletic skills of blacks
vs. whites, 80; Bernier on, 52–53, 55;
blood tests for determining, 63–65, 81,
126–27; Blumenbach on, 61, 65; Buffon
on, 60–61, 65; "chimpanzee" as a racial
insult, 69–71, *71*, 293; and civilization,
76–77; colloquial vs. scientific sense of,
62; Coon on, 76–77; cultural bases of
classifications of, 78–80, 83, 93–94, 97–
98 (*see also* folk heredity); and cultural
diversity, 87–88; and economics/politics
of Europe, 54–55; and evolutionary descent, patterns of, 58–59; and facial features, 51; genetic basis for, 81–82; and
God's bounty vs. parsimony, 52; and
head size/shape, 78–80; *Homo,* kinds of,
55; and the Human Genome Diversity
Project, 200–202, *201*, 209–12, 216;
identification with racial groups, 94;
and Kennewick Man, 233–36; Linnaeus
on, 55–61, 57 (table); and miscegenation
laws, 68; and the Mitochondrial Eve report, 83; Noah's Sons model of, 53–54,
67, 97–98, 211–12; and physical anthropology as Linnaean science, 61–69; and
populations, 74–75, 77, 80–82, 132–34;
racial vs. biological heredity, 68–69; as a
scientific organizing principle, 52–61, 57
(table); and skull form, 236–37; social
problems associated with, 94–95; social
significance of, 243; and taxonomism,
136–38. *See also* human variation, meaning of
racial anthropology, Nazi vs. American,
75, 140
racism, 139–46, 210
RAFI (Rural Advancement Foundation
International), 206, 213
Ramapithecus, 11–12
Rassenhygiene, 277
rationalism, 274–75, 276
relatives/kinship, 243, 249–51
relativism, 227, 272–74
religion: the Bible, 246–47, 254–55; Christianity, 140; God's bounty vs. parsimony, 52; human–chimpanzee shared
ancestry as threatening to, 254–55, 257;

and science, 6, 274, 276. *See also* creationism
"Report to an Academy" (Kafka), 40
reproduction vs. sexuality, 110–11
reproductive rates, human vs. ape, 194–95
reptiles and birds, 44–45
Retzius, Anders, 89
rights. *See* apes, human rights for; human rights
Roberts, Leslie, 202, 203
Roosevelt, Theodore, 119, 268–69
Rorvick, David: *In His Image,* 223
Rousseau, Jean-Jacques, 172, 175, 176, 272–73; *Discourse on the Origin of Inequality,* 168; *The Social Contract,* 168
Rural Advancement Foundation International (RAFI), 206, 213
Rushton, J. Philippe, 79

Sagan, Carl, 274
SAIIC (South-and-Meso American Indian Information Center), 207
Sarich, Vincent, 10–13, 261–62
satyr, 15–17, *16*
Sawyer, Forrest, 175
Schiebinger, Londa, 50
Schneider, David, 135, 249
Schrödinger, Erwin, 280
Schultz, Adolph, 70–71, *71,* 293
science, 266–88; anthropological study of, 266–67; and anthropology, 286–88; authority vs. accuracy in, 278–79; authority of/responsibility in, 119–22, 148–49, 157–58, 221, 222, 278–79, 284–88; burden of proof in, 126, 144, 245; as concerned with the natural rather than the supernatural, 257–58; ethnocentrism of, and the future of, 266, 284–88; and the eugenics movement, 268–71, 277; and facts, 267–68, 270–72; vs. folk knowledge, 2, 5, 93; fundamentalist Christian subversion of education in, 185, 255–56, 280; and humanism, 284, 285, 286, 288; vs. humanities, 1; and human rights, 269–70; and meaninglessness of life, 281–83; and myth, 272–73, 280–84; and nature, mastery of, 274;

and new knowledge, 254; rationalism of, 274–75, 276; as rational and social, 277–78; and relativism, 272–74; vs. religion, 6, 274, 276; standpoints within, 4; vs. superstition, 276; and technology vs. mythology, 272–73; testability in, 174; vs. transcendentalism, 274–75
Science, 113; on the Human Genome Diversity Project, 202, 203, 204, 206, 216
scientific classification. *See* classification
Scopes (John T.) trial, 42, 255, 257, 258–59, 269
Seattle Times, 229, 231
Second Treatise on Civil Government (Locke), 167–68
segregation, 77
self-discovery, 224–25
sexism, 140
sexuality, 110–11, 175–76, 188. *See also* homosexuality
Shapiro, Harry, 91
Shepard, Glenn, 155
Shockley, William, 210
Show Boat, 243
Silk Stockings, 275
Simpson, George Gaylord, 21, 42–43
Singer, Peter: *The Great Ape Project,* 186, 189
skull size/shape, 78–80, 89–91, 177, 236–37
slavery, 76
Sleeper (Allen), 223
Small, Meredith, 156
Smitten, 17–18
Snow, C. P., 1
The Social Contract (Rousseau), 168
social Darwinism, 130–31, 259–61
sociobiology, 169–70
Sokal, Alan, 268
Sollas, William J.: *Ancient Hunters,* 260
South-and-Meso American Indian Information Center (SAIIC), 207
Soviet Revolution, 277
Speck, Richard, 124
Spencer, Baldwin, 169
Spencer, Herbert, 161
Sputnik, 256
Stanford, Craig, 176

Stanford, Dennis, 232, 233
Steele, Gentry, 232
sterilization, involuntary, 119–20, 121, 269, 271
Stewart, Patrick, 234–35
Stoneking, Mark, 83
Sunset Boulevard, 187
superstition, 276
Sykeses, Y chromosomes of, 247, 301
Systema naturae (System of Nature; Linnaeus), 20–21, 49–50, 55–56, 57 (table)

taxonomism, 136–38
technology, 233–34, 272–73
Third Reich. *See* Nazis
Tierney, Patrick, 172; *Darkness in El Dorado,* 287
Time, 209–10, 211
time-keeping systems, 277
Tishkoff, Sarah, 84, 85
transcendentalism, 274–75
Travels of Sir John Mandeville (Mandeville), 166
Tri-City Herald, 228, 229
Trobriand Islanders (Melanesia), 129
Tsonga people (southern Africa), 129
Tulp, Nicolaas, 15–17, *16*
Twain, Mark, 30
twin studies, 114–15, 149–53, 158
Tylor, Edward Burnett, 181, 287
Tyson, Edward, 17, *18,* 264

Udall, Mo, 227
Umatilla people (Washington State), 231, 238–39
UNESCO, 169

Vershuer, Otmarr Freiherr von, 270
violence, 70–71, 107–8, 293. *See also* aggression

waist-to-hip ratio of women, 154–55
Walsh, Bruce, 199
Washburn, Sherwood, 73, 74–75, 77
Watson, James, 148, 297
wet-nursing, 50
Wilson, Allan, 10–12, 32–33, 83, 86; on the Human Genome Diversity Project, 202, 203
Wilson, E. O., 108, 274, 276, 278
Wolf, Eric, 170
women's figures, 154–55
World Council of Indigenous Peoples (Guatemala), 207
Wrangham, Richard, 159, 172–73, 174
Wright, Lawrence, 149, 151, 152
Wright, Sewall, 251

xenophobia, 141–42

Yanomamo people (Venezuela), 171–72, 179, 287
Yu, Douglas, 155

Zuckerkandl, Emile, 42

Indexer:	Carol Roberts
Compositor:	Binghamton Valley Composition, LLC
Text:	12/14.5 Adobe Garamond
Display:	Perpetua and Adobe Garamond
Printer and binder:	Maple-Vail Manufacturing Group